T0283655

The Prisoner's Dilemma

The Prisoner's Dilemma is one of the most fiercely debated thought experiments in philosophy and the social sciences, presenting the simple insight that when two or more agents interact, the actions that most benefit each individual may not benefit the group. The fact that when you do what is best for you, and I do what is best for me, we end up in a situation that is worse for both of us makes the Prisoner's Dilemma relevant to a broad range of everyday phenomena. This volume of new essays from leading philosophers, game theorists, and economists examines the ramifications of the Prisoner's Dilemma, the directions in which it continues to lead us, and its links to a variety of topics in philosophy, political science, social science, economics, and evolutionary biology. The volume will be a vital and accessible resource for upper-level students as well as for academic researchers.

Martin Peterson is Sue and Harry E. Bovay Professor in the Department of Philosophy at Texas A&M University. His recent publications include *Non-Bayesian Decision Theory* (2008), *An Introduction to Decision Theory* (Cambridge, 2009), and *The Dimensions of Consequentialism* (Cambridge, 2013).

Classic Philosophical Arguments

Over the centuries, a number of individual arguments have formed a crucial part of philosophical enquiry. The volumes in this series examine these arguments, looking at the ramifications and applications which they have come to have, the challenges which they have encountered, and the ways in which they have stood the test of time.

Title in the series

The Prisoner's Dilemma
Edited by Martin Peterson

Forthcoming title

The Original Position
Edited by Timothy Hinton

The Prisoner's Dilemma

Edited by
Martin Peterson

Shaftesbury Road, Cambridge CB2 8EA, United Kingdom

One Liberty Plaza, 20th Floor, New York, NY 10006, USA

477 Williamstown Road, Port Melbourne, VIC 3207, Australia

314–321, 3rd Floor, Plot 3, Splendor Forum, Jasola District Centre, New Delhi – 110025, India

103 Penang Road, #05–06/07, Visioncrest Commercial, Singapore 238467

Cambridge University Press is part of Cambridge University Press & Assessment, a department of the University of Cambridge.

We share the University's mission to contribute to society through the pursuit of education, learning and research at the highest international levels of excellence.

www.cambridge.org
Information on this title: www.cambridge.org/9781107044357

First published 2015

A catalogue record for this publication is available from the British Library

Library of Congress Cataloging-in-Publication data
The prisoner's dilemma / edited by Martin Peterson.
pages cm
Includes bibliographical references and index.
ISBN 978-1-107-04435-7 (Hardback) – ISBN 978-1-107-62147-3 (Paperback)
1. Game theory. 2. Prisoner's dilemma game. 3. Cooperativeness–Mathematical models.
I. Peterson, Martin, 1975–
HB144.P75 2015
519.3–dc23 2014050242

ISBN 978-1-107-04435-7 Hardback
ISBN 978-1-107-62147-3 Paperback

Contents

Contributors

Anna Alexandrova is Lecturer in the Department of History and Philosophy of Science, Cambridge University

Jeffrey A. Barrett is Professor and Chancellor's Fellow in the Department of Logic and Philosophy of Science, University of California, Irvine

José Luis Bermúdez is Associate Provost and Professor of Philosophy, Texas A&M University

Cristina Bicchieri is Sascha Jane Patterson Harvie Professor of Social Thought and Comparative Ethics, University of Pennsylvania

Ken Binmore is Professor of Economics, University of Bristol

Giacomo Bonanno is Professor of Economics, University of California, Davis

Luc Bovens is Professor of Philosophy, London School of Economics and Political Science

Geoffrey Brennan is Distinguished Research Professor, University of North Carolina at Chapel Hill and Australian National University

Michael Brooks is Associate Professor in the School of Business and Economics, University of Tasmania

David Gauthier is Distinguished Service Professor Emeritus, University of Pittsburgh

Daniel M. Hausman is Herbert A. Simon and Hilldale Professor of Philosophy, University of Wisconsin-Madison

Charles Holt is A. Willis Robertson Professor of Political Economy, University of Virginia

Cathleen Johnson is Lecturer in Philosophy, Politics, Economics and Law, University of Arizona

Douglas MacLean is Professor of Philosophy, University of North Carolina at Chapel Hill

Robert Northcott is Senior Lecturer in Philosophy, University of London, Birkbeck

Martin Peterson is Sue and Harry Bovay Professor in the Department of Philosophy, Texas A&M University

David Schmidtz is Kendrick Professor of Philosophy, University of Arizona

Alessandro Sontuoso is Postdoctoral Researcher in the Philosophy, Politics and Economics Program, University of Pennsylvania

Paul Weirich is Curators' Professor of Philosophy, University of Missouri-Columbia

Introduction

Martin Peterson

0.1 An ingenuous example

The Prisoner's Dilemma is one of the most fiercely debated thought experiments in philosophy and the social sciences. Unlike many other intellectual puzzles discussed by academics, the Prisoner's Dilemma is also a type of situation that many of us *actually encounter* in real life from time to time. Events as diverse as traffic jams, political power struggles, and global warming can be analyzed as Prisoner's Dilemmas.

Albert W. Tucker coined the term "Prisoner's Dilemma" during a lecture in 1950 in which he discussed the work of his graduate student John F. Nash.[1] Notably, Nash won the Nobel Prize in Economics in 1994 and is the subject of the Hollywood film *A Beautiful Mind* (which won four Academy Awards). If this is the first time you have come across the Prisoner's Dilemma, I ask you to keep in mind that the following somewhat artificial example is just meant to illustrate a much more general phenomenon:

> Two gangsters, Row and Col, have been arrested for a serious crime. The district attorney gives them one hour to either confess or deny the charges. The district attorney, who took a course in game theory at university, explains that if both prisoners confess, each will be sentenced to ten years in prison. However, if one confesses and the other denies the charges, then the prisoner who confesses will be rewarded and get away with serving just one year. The other prisoner will get twenty years. Finally, if both prisoners deny the charges, each will be sentenced to two years. The prisoners are kept in separate rooms and are not allowed

[1] Nash writes: "It was actually my thesis adviser, who dealt with my thesis in preparation (on 'Non-Cooperative Games') who introduced the popular name and concept by speaking at a seminar at Stanford U. (I think it was there) while he was on an academic leave from Princeton. And of course this linked with the fact that he was naturally exposed to the ideas in 'Non-Cooperative Games' (my thesis, in its published form)." (Email to the author, December 14, 2012.)

		COL	
		Deny	Confess
ROW	Deny	−2, −2	−20, −1
	Confess	−1, −20	−10, −10

Figure 0.1 The Prisoner's Dilemma

to communicate with each other. (The significance of these assumptions will be discussed at the end of this section.)

The numbers in Figure 0.1 represent each prisoner's evaluation of the four possible outcomes. The numbers −1, −20 mean one year in prison for Row and twenty years for Col, and so on. Naturally, both prisoners prefer to spend as little time in prison as possible.

The Prisoner's Dilemma has attracted so much attention in the academic literature because it seems to capture something important about a broad range of phenomena.[2] Tucker's story is just a colorful illustration of a general point. In order to understand this general point, note that both Row and Col are rationally required to confess their crimes, *no matter what the other player decides to do.* Here is why: If Col confesses, then ten years in prison for Row is better than twenty; and if Col denies the charges, then one year in prison is better for Row than two. By reasoning in analogous ways we see that Col is also better off confessing, regardless of what Row decides to do. This is somewhat counterintuitive, because both prisoners know it would be better for both of them to deny the charges. If Row and Col were to deny the charges, they would each get just two years, which is better than ten. The problem is that as long as both prisoners are fully rational, there seems to be no way for them to reach this intuitively plausible conclusion.

The general lesson is that whenever two or more players interact and their preferences have a very common and reasonable structure, the actions that most benefit each individual *do not benefit the group.* This makes the Prisoner's Dilemma relevant to a broad range of social phenomena. When I do what is best for me, and you do what is best for you, we end up in a situation that is *worse for both of us.* The story of the two prisoners is just a tool for illustrating this point in a precise manner.

[2] Note, however, that some scholars think this attention is unwarranted; see Binmore's contribution to this volume.

We cannot avoid the Dilemma, at least not in a straightforward way, by allowing the prisoners to communicate and coordinate their actions. If Col and Row each promises the other that he will deny the charges, it would still be rational for both men to confess, given that the numbers in Figure 0.1 represent *everything* that is important to them. When the district attorney asks the players to confess, they no longer have a rational reason to keep their promises. If Row confesses and Col does not, then Row will get just one year, which is better than two. It is also better for Row to confess if Col confesses. Therefore, it is better for Row to confess *irrespective of* what Col does. And because the game is symmetric, Col should reason exactly like Row and confess too.

If keeping a promise is considered to be valuable for its own sake, or if a prisoner could be punished for not keeping a promise, then the structure of the game would be different. By definition, such a modified game would no longer qualify as a Prisoner's Dilemma. These alternative games, also studied by game theorists, are less interesting from a theoretical point of view. In this volume, the term "Prisoner's Dilemma" refers to any game that is structurally equivalent to that depicted in Figure 0.1.[3]

For an alternative and perhaps more realistic illustration of the Prisoner's Dilemma, consider two competing car manufacturers: Row Cars and Col Motors. Each company has to decide whether to sell their cars for a high price and make a large profit from each car sold, or lower the price and sell many more vehicles with a lower profit margin. Each company's total profit will depend on whether *the other company* decides to set its prices high or low. If both manufacturers sell their cars at high prices, each will make a profit of $100 million. However, if one company opts for a low price and the other for a high price, then the latter company will sell just enough cars to cover its production costs, meaning that the profit will be $0. In this case, the other company will then sell many more cars and make a profit of $150 million. Finally, if both manufacturers sell their cars at low prices, they will sell an equal number of cars but make a profit of only $20 million. See Figure 0.2.

Imagine that you serve on the board of Row Cars. In a board meeting you point out that *irrespective of what Col Motors decides to do*, it will be better for your company to opt for low prices. This is because if Col Motors sets its price low, then a profit of $20M is better than $0; and if Col Motors sets its price high, then a profit of $150M is better than $100M. Moreover, because

[3] Note that a more precise definition is stated in Section 0.2 of this chapter.

		Col Motors	
		High Price	Low Price
Row Cars	High Price	\$100M, \$100M	\$0M, \$150M
	Low Price	\$150M, \$0M	\$20M, \$20M

Figure 0.2 Another illustration of the Prisoner's Dilemma

the game is symmetric, Col Motors will reason in the same way and also set a low price. Therefore, both companies will end up making a profit of \$20M each, instead of \$100M.

The conclusion that the two companies will, if rational, opt for low prices is not something we have reason to regret. Not all Prisoner's Dilemmas are bad for ordinary consumers. However, for Row Cars and Col Motors it is no doubt unfortunate that they are facing a Prisoner's Dilemma. If both companies could have reached a *binding agreement* to go for high prices, both companies would have made much larger profits (\$100M). This might explain why government authorities, in protecting consumers' interests, do their best to prevent cartels and other types of binding agreements about pricing.[4]

0.2 Some technical terms explained

Let us try to formulate the Prisoner's Dilemma using a more precise vocabulary. Consider Figure 0.3. By definition, the game depicted in this figure is a Prisoner's Dilemma if outcome A is preferred to B, and B is preferred to C, and C is preferred to D. (That is, $A > B > C > D$.) For technical reasons, we also assume that $B > (A + D) / 2$.[5]

In its classic form, the Prisoner's Dilemma is a *two-player, non-cooperative, symmetric, simultaneous-move* game that has only one *Nash equilibrium*. The italicized terms in the foregoing sentence are technical terms with very precise meanings in game theory.

[4] Some scholars question this type of explanation; see the contribution by Northcott and Alexandrova for a detailed discussion.

[5] This assumption is needed for ensuring that the players cannot benefit more from alternating between cooperative and non-cooperative moves in repeated games, compared to playing mutually cooperative strategies. (Note that we presuppose that the capital letters denote some cardinal utilities. Otherwise the mathematical operations would be meaningless.)

		COL	
		Cooperate	Do not
ROW	Cooperate	B, B	D, A
	Do not	A, D	C, C

Figure 0.3 The generic Prisoner's Dilemma

A *two-player* game is a game with exactly two players. Many Prisoner's Dilemmas are two-player games, but some have three, one hundred, or *n* players. Consider global warming, for instance. I prefer to emit a lot of carbon dioxide irrespective of what others do (because this enables me to maintain my affluent lifestyle), but when all *n* individuals on the planet emit huge amounts of carbon dioxide, because this is the best strategy for each individual, that leads to global warming and other severe problems for all of us.

A *non-cooperative* game is a game in which the players are unable to form binding agreements about what to do. Whether the players actually cooperate or not is irrelevant. Even if the players promise to cooperate with each other, the game would still be a non-cooperative game as long as there is no mechanism in place that forces the players to stick to their agreements. In a non-cooperative game, the players can ignore whatever agreement they have reached without being punished.

That the Prisoner's Dilemma is a *symmetric* game just means that all players are faced with the same set of strategies and outcomes, meaning that the identity of the players is irrelevant. Symmetric games are often easier to study from a mathematical point of view than non-symmetric ones.

That a game is a *simultaneous-move* game means that each player makes her choice without knowing what the other player(s) will do. It is thus not essential that the players make their moves at exactly the same point in time. If you decide today what you will do tomorrow without informing me, the game will still be a simultaneous-move game as long as I also make my move without informing you about it in advance.

The Prisoner's Dilemma is sometimes a simultaneous-move game, but it can also be stated as a *sequential* game in which one player announces his move before the other. Figure 0.4 illustrates a sequential Prisoner's Dilemma in which Player 1 first chooses between two strategies C ("cooperate") and D ("defect"), which is followed by Player 2's choice. The outcome (A1, A2) means that Player 1 gets something worth A1 to him and Player 2 gets A2 units of value. As long as $A1 > B1 > C1 > D1$ and $A2 > B2 > C2 > D2$

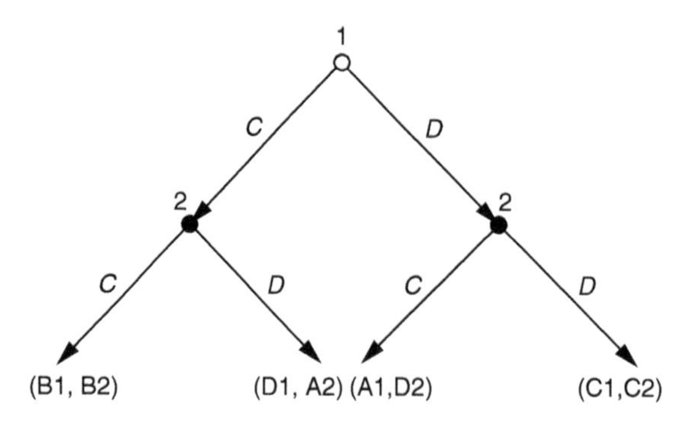

Figure 0.4 In a sequential Prisoner's Dilemma it holds that A1 > B1 > C1 > D1 and A2 > B2 > C2 > D2

the dominance reasoning that drives single-shot versions of the Prisoner's Dilemma will go through.

As explained above, both players' non-cooperative strategies dominate their cooperative strategies in a Prisoner's Dilemma. This type of game therefore has only one *Nash equilibrium*. A Nash equilibrium is a situation in which no player has any reason to unilaterally switch to any other strategy. In his doctoral dissertation, Nash defined the equilibrium concept that bears his name in the following way:

> An equilibrium point is [a set of strategies] such that each player's . . . strategy maximizes his pay-off if the strategies of the others are held fixed. Thus each player's strategy is optimal against those of the others. (Nash 1950: 3)

The key insight is that a pair of strategies is in equilibrium just in case it holds true that *if* these strategies are chosen by the players, then none of the players could reach a better outcome by *unilaterally* switching to another strategy. In other words, rational players will do whatever they can to ensure that they do not feel *unnecessarily* unhappy about their decision, meaning that if a player could have reached a better outcome by *unilaterally* switching to another strategy, the player would have done so.

In order to understand why Nash equilibria are important in game theory, imagine that you somehow knew that your opponent was going to play the cooperative strategy. Would you then, if you were to also play the cooperative strategy, have a reason to unilaterally switch to the non-cooperative strategy ("defect")? The answer is yes, because you would actually gain more by doing so. This shows that if we take Nash's equilibrium concept to be a necessary condition for a plausible principle for how to tackle the Prisoner's Dilemma, then we cannot expect rational players to cooperate.

0.3 The repeated Prisoner's Dilemma

The analysis of the Prisoner's Dilemma depends crucially on how many times it is played. In the single-shot Prisoner's Dilemma a rational player will play the non-cooperative strategy ("defect", i.e. confess the crime), because this strategy dominates the cooperative strategy. Binmore claims in his contribution to this volume that this "is trivial and entirely unworthy of the attention that has been devoted to it."[6] Although Binmore's claim will perhaps not be accepted by everyone, it is important to realize that he is right that it does indeed make a huge difference whether the Prisoner's Dilemma is played only once or many times.

Let us first consider what happens if the Prisoner's Dilemma is repeated (or "iterated") a *finite* number of times. Suppose, for instance, that the Prisoner's Dilemma is repeated exactly three times and that the players know this. To make the example a bit realistic, imagine that each player is a car manufacturer who has exactly three opportunities to adjust the price of some car model during its lifespan. In each round, the companies can choose between a high and a low price. See Figure 0.2. The two players can then reason backwards: In the third round they will know that they are playing the last round; they therefore have no reason to cooperate in that round, meaning that the third and last round can be analyzed as a single-shot Prisoner's Dilemma; each player's non-cooperative strategy dominates her cooperative strategy, so each player will refrain from cooperating in the last round and set a low price.

When the players get to the penultimate round, both players know that neither of them will cooperate in the last round. They therefore have no incentive to cooperate in the penultimate round, because they have no reason to believe that their current behavior will increase the likelihood that the other player will cooperate in the future. The penultimate round can hence be analyzed in the same way as the last one: both players will set their prices low. Finally, the first round can be analyzed exactly like the penultimate round. The players have no reason to cooperate in this round because there is no reason to think their current actions will increase the likelihood that the other players will cooperate in the future.

The argument summarized above is known as the *backward induction argument*. The basic idea is that rational players should reason backwards, from the last round to the first. Note that from a logical point of view it makes no difference if we apply this argument to a Prisoner's Dilemma that is

[6] See Binmore's contribution to this volume, p. 17.

repeated two, three, or a thousand times. However, it is worth keeping in mind that the larger the number of rounds is, the more unintuitive and descriptively implausible the argument will become. Experimental results show that if people get to play a large (but finite) number of rounds, they are likely to cooperate with their opponent because they think this will be rewarded by the opponent in future rounds of the game.[7] This is also how we would expect real-world car manufactures to reason. When they play against each other, they have no pre-defined last round in mind from which they reason backwards. On the contrary, car manufacturers and other large corporations typically seem to think the game they are playing is likely to be repeated in the future, which makes it rational to take into account how one's opponent might respond in the next round to the strategy one is playing now. This indicates that the backward induction argument is a poor analysis of many repeated Prisoner's Dilemmas in the real world.

Many scholars consider the *indefinitely* repeated Prisoner's Dilemma to be the most interesting version of the game. The indefinitely repeated Prisoner's Dilemma need not be repeated infinitely many times. What makes the game indefinitely repeated is the fact that there is no point at which the players *know in advance* that the next round will be the last. The key difference between finitely repeated and indefinitely repeated versions of the Prisoner's Dilemma is thus not how many times the game is actually played, but rather what the players know about the future rounds of the game. Every time the indefinitely repeated game is played, the players know that there is some *non-zero probability that the game will be played again* against the same opponent. However, there is no pre-defined and publicly known last round of the game. Therefore, the backward induction argument cannot get off the ground, simply because there is no point in time at which the players know that the next round will be the last.

So how should rational players behave in the indefinitely repeated Prisoner's Dilemma? The key insight is that each player has reason to take the future behavior of the opponent into account. In particular, there is a risk that your opponent will punish you in the future if you do not cooperate in the current round. In "the shadow of the future," you therefore have reason to cooperate. Imagine, for instance, that your opponent has played cooperative moves in the past. It then seems reasonable to conclude that your opponent is likely to cooperate next time too. To keep things simple, we assume that your opponent has cooperated in the past *because* you have

[7] See Chapter 13.

played cooperative moves in the past. Then it seems foolish to jeopardize this mutually beneficial cooperation by playing the dominant strategy in the current round. If you do so, your opponent will probably not cooperate in the next round. It is therefore better for you to cooperate, all things considered, despite the fact that you would actually be better off in this round by not cooperating.

In the past thirty years or so, game theorists have devoted much attention to indefinitely repeated Prisoner's Dilemmas. The most famous strategy for these games is called *Tit-for-Tat*. Players who play Tit-for-Tat always cooperate in the first round, and thereafter adjust their behavior to whatever the opponent did in the previous round. Computer simulations, as well as theoretical results, show that Tit-for-Tat does at least as well or better than nearly all alternative strategies. Several contributions to this volume discuss the indefinitely repeated version of the Prisoner's Dilemma.

0.4 Overview

This volume comprises fourteen new essays on the Prisoner's Dilemma. The first three chapters set the stage. The next ten chapters zoom in on a number of specific aspects of the Dilemma. The final chapter draws conclusions.

In Chapter 1, Ken Binmore defends two claims. First, he argues that all arguments for cooperating in the single-shot Prisoner's Dilemma proposed in the literature so far are fallacious. Such arguments either alter the structure of the game, or introduce additional, questionable assumptions that we have no reason to accept. The only rational strategy in the non-cooperative version of the game is, therefore, the single-shot strategy. Binmore's second claim concerns the connection between the Prisoner's Dilemma and Kant's categorical imperative. Binmore's conclusion is that although the indefinitely repeated Prisoner's Dilemma can shed light on the evolution of a number of social norms, "the Prisoner's Dilemma shows that [the categorical imperative] can't be defended purely by appeals to rationality as Kant claims." So, according to Binmore, Kant was wrong, at least if the notion of rationality he had in mind was the same as that researched by game theorists.

David Gauthier, in Chapter 2, disagrees with Binmore on several issues. The most important disagreement concerns the analysis of single-shot Prisoner's Dilemmas. Unlike Binmore, Gauthier claims that it is sometimes (but not always) rational to cooperate in a single-shot Prisoner's Dilemma. The argument for this unorthodox view goes back to his earlier work. As Gauthier now puts it, the main premise is that, "if cooperation is, and is recognized by (most) other persons to be possible and desirable, then it is rational for each to be a cooperator." Why? Because each person's own objectives can be more easily realized if they agree to bring about a mutually beneficial outcome. It is therefore rational for each person to think of herself as a cooperator and deliberate together with the other player about what to do.

It should be stressed that Gauthier's argument is based on a theory of practical rationality that differs in important respects from the one usually employed by game theorists. According to the traditional theory, an agent is rational if and only if she can be described as an expected utility maximizer. This, in turn, means that her preferences obey a set of structural conditions (called "completeness," "transitivity," "independence," and "continuity"). It is moreover assumed that what agents prefer is "revealed" in choices, meaning that the agent always prefers what she chooses. Gauthier rejects this traditional picture and sketches an alternative. In Gauthier's view, preference and choice are separate entities, meaning that an agent can make choices that conflict with her preferences. This alternative account of practical rationality, which will probably appeal to many contemporary philosophers, is essential for getting Gauthier's argument off the ground. In order to coherently defend the claim that it is sometimes rational to cooperate in the single-shot Prisoner's Dilemma, the preferences that constitute the game cannot be identified with the corresponding choices.

In Chapter 3, Daniel M. Hausman discusses the notion of preference in the Prisoner's Dilemma. Economists take the player's preference to incorporate all the factors other than beliefs that determine choices. A consequence of this close relation between preference and choice is that it becomes difficult to separate the influence of moral reasons and social norms from other factors that may influence one's choice. If you choose to cooperate with your opponent in what appears to be a Prisoner's Dilemma because you feel you are under a moral obligation to do so, or because your behavior is influenced by some social norm, then you prefer to cooperate, and the game you are playing is not a Prisoner's Dilemma. If, as Hausman believes, cooperation in what appear to be Prisoner's Dilemmas often shows that people are not in fact playing Prisoner's Dilemmas, then the experimental results pose no challenge to the simple orthodox analysis of the Prisoner's Dilemma. But this leaves game theorists with the task of determining what game individuals in various strategic situations are actually playing. Hausman discusses some sophisticated ways of thinking about how to map "game forms" into games, especially in the work of Amartya Sen and Cristina Bicchieri, but he concludes that there is a great deal more to be done.

In Chapter 4, Robert Northcott and Anna Alexandrova criticize the explanatory power of the Prisoner's Dilemma. They point out that more than 16,000 academic articles have been published about the Prisoner's Dilemma over the past forty years. However, the empirical payoff from this enormous scholarly investment is underwhelming. A great deal of diverse phenomena have been claimed to instantiate the Prisoner's Dilemma, but when looked at closely, Northcott and Alexandrova argue, most of these applications are too casual to be taken seriously as explanations. On the rare occasions when the model is applied with proper care, its value as a causal explanation is actually dubious. In support of this claim, they discuss as a case study the "live-and-let-live" system in the World War I trenches that Robert Axelrod famously claimed was an instance of the indefinitely iterated Prisoner's Dilemma played with the Tit-for-Tat strategy. Northcott and Alexandrova question the predictive, explanatory, and even heuristic power of this account. The historical record undermines its analysis of the co-operation in the trenches, and none of the standard defenses of idealized modeling in economics can save it. They conclude that the value of the Prisoner's Dilemma for social science has been greatly exaggerated.

The next two chapters, Chapters 5 and 6, discuss the effect of communication on cooperation. Several experimental studies show that if the players are able to communicate with each other, they are more likely to cooperate. Jeffrey A. Barrett discusses how evolutionary game theorists can explain this. Barrett focuses on a particular sub-class of evolutionary games, namely Prisoner's Dilemma games with pre-play signaling. The term "pre-play signaling" means that one or more players are able to send signals to the other players before the game is played. At the beginning of the evolutionary process the senders send out random signals, which the receiver uses for making random predictions about the sender's dispositions. However, as the game is repeated many times, the senders and receivers learn from experience to interpret the various signals and make almost entirely correct descriptive and predictive interpretations of the other player's disposition. This evolutionary process can help to explain why people often seem to cooperate much more than they should, according to more orthodox theories, in Prisoner's Dilemma-like games in which there is no pre-play signaling.

Cristina Bicchieri and Alessandro Sontuoso also discuss the effect of communication on cooperation, but from a different perspective. Their focus is on the role of social norms. In her earlier work, Bicchieri has argued that when agents face unfamiliar situations, communication can help the players focus on social norms so as to reach solutions that are, under certain conditions, beneficial to the players. In the present contribution, Bicchieri's and Sontuoso's main concern is with a type of social dilemmas called sequential trust games. They present a general theory of norm conformity, and provide a novel application illustrating how a framework that allows for different conjectures about norms is able to capture what they call the "focusing function" of communication.

In Chapter 7, José Luis Bermúdez breathes new life into a debate initiated by David Lewis. According to Lewis, the Prisoner's Dilemma and Newcomb's Problem are one and the same problem, with the non-cooperation strategy in the Prisoner's Dilemma essentially a notational variant of one-boxing in Newcomb's Problem – in both cases the apparent appeal of the other strategy is due to neglecting the causal dimension of the decision problem. This is important to Lewis in the context of his arguments for causal decision theory, because Prisoner's Dilemmas are much more widespread than Newcomb Problems and so it becomes correspondingly more plausible that the causal dimension of decision problems always needs to be taken into account, as Lewis and other causal decision theorists claim. According to Bermúdez, however, Lewis's argument is indirectly self-defeating. His argument works only when a player believes that the other player will essentially behave as she does – and yet, Bermúdez suggests, this effectively transforms the game so that it is no longer a Prisoner's Dilemma. As Bermúdez points out, this result is not very surprising. The Prisoner's Dilemma is a two-person game and, although Newcomb's Problem does involve two individuals, they are not interacting game-theoretically. Newcomb's Problem is a problem of parametric choice, not of strategic choice, and it would be surprising indeed if the Prisoner's Dilemma, standardly understood as a strategic interaction, turned out to be equivalent to a pair of simultaneous parametric choices.

Giacomo Bonanno analyzes the role of conditionals in the Prisoner's Dilemma. He begins Chapter 8 with the reminder that there are two main groups of conditionals: indicative conditionals ("If Oswald did not kill Kennedy, then someone else did") and counterfactual conditionals ("If Oswald had not shot Kennedy, then someone else would have"). Bonanno's aim is to clarify the role of counterfactual reasoning in the Prisoner's Dilemma, and in particular in arguments that seek to show that it is rational to cooperate in single-shot versions of this game. According to Bonanno, such arguments depend on beliefs of the following type: "If I cooperate then my opponent will also cooperate, but if I do not cooperate then my opponent will refrain from doing so. Therefore, if I cooperate I will be better off than if

I do not. Hence, it is rational for me to cooperate." Bonanno argues that beliefs of this type might sometimes be implausible or far-fetched, but they are not necessarily irrational. If true, this indicates that it can be rational to cooperate in single-shot Prisoner's Dilemmas, contrary to what many scholars currently think.

The next two chapters discuss the relation between the two-person Prisoner's Dilemma and many-person versions of the Dilemma. In Chapter 9, Luc Bovens investigates the relation between the n-players Prisoner's Dilemma and the Tragedy of the Commons, introduced in a famous article in *Science* by Garrett Hardin in 1968. The Tragedy of the Commons refers to a situation in which multiple independent and self-interested players are rationally obligated to deplete a shared and limited resource, even when all players know that it is not in their long-term interest to do so. Consider, for instance, issues related to sustainability. It is sometimes rational for each self-interested individual to keep polluting the environment, even when he or she knows that this will leads to a situation that is suboptimal for everyone. Many authors have argued that this tragic insight is best understood as an n-person Prisoner's Dilemma, but Bovens questions this received view. His claim is that the Tragedy of the Commons should rather be modeled as what he calls a Voting Game. In Bovens' terminology, a Voting Game is a game in which you rationally prefer to cast a vote just in case you belong to a minimal winning coalition, but otherwise prefer not to cast a vote, because then your action has no positive effect.

In Chapter 10, Geoffrey Brennan and Michael Brooks discuss the relation between the two-person Prisoner's Dilemma and the n-person versions of the Dilemma in more detail. The consensus view in the economic literature is that there are very considerable differences between the two-person case and n-person cases. It has, for instance, been suggested that each player's individual benefit of being selfish is typically smaller in the two-person Prisoner's Dilemma than in n-person games. Brennan and Brooks scrutinize this and other similar claims about the impact of the group size. Their conclusion is that the significance of the group size is much more complex than what has been previously recognized.

My own contribution, in Chapter 11, explores an argument for cooperating in the single-shot two-person Prisoner's Dilemma. This argument is based on some ideas originally proposed by Plato and Gregory Kavka. The key premise is that rather than thinking of each player as someone who assigns utilities to outcomes, we should (at least sometimes) think of the players as a set of subagents playing games against other subagents.

Consider, for instance, your decision to eat healthy or unhealthy food. Your health-oriented subagent ranks dishes that are healthy but bland above those that are unhealthy but flavorful. Your taste-oriented subagent has the opposite preference ordering. Do you choose the bland but healthy salad or the delicious but not-so-healthy burger? What you eventually end up doing depends on the outcome of the "inner struggle" between your subagents. This struggle can be described as an *internal* Prisoner's Dilemma if the preference orderings of the subagents fulfill some plausible assumptions. I argue that in an indefinitely repeated inner struggle between your subagents, the rational strategy will be to cooperate even if the game you yourself face is a single-shot Prisoner's Dilemma.

In Chapter 12, Douglas MacLean discusses how the Prisoner's Dilemma can improve our understanding of climate change. His point of departure is the tragic observation that, "after several decades of increased understanding about the causes and effects of climate change, the global community has failed to take any meaningful action in response." MacLean notes that the game theoretical structure that is most relevant for analyzing this phenomenon is the Tragedy of the Commons. However, unlike Bovens, he does not take a stand regarding whether the Tragedy of the Commons is just an *n*-person Prisoner's Dilemma or some other, slightly different type of game. MacLean also emphasizes the ethical dimensions of climate change.

Chapter 13, by David Schmidtz, Cathleen Johnson, and Charles Holt, presents new experimental results about how people actually behave in Prisoner's Dilemma-like games. They offer an extensive overview of the experimental literature and then go on to present a new experiment in which eighty-four subjects played a finitely repeated Prisoner's Dilemma with a publicly known number of rounds. As explained above, in such a game it is not rational to cooperate in the last round, and because of the backwards induction argument, it is irrational to cooperate in all rounds of the game. However, despite this, Schmidtz, Johnson, and Holt find that players cooperate to a high degree in the first rounds of the game, but then defect quite significantly in the last round. This is consistent with earlier findings, but they also add some new results. The most important finding is the observation that the scale of interaction among the players matters. If people interact more with each other (or, in technical jargon, if the "link duration" is long) then this yields a higher rate of cooperation.

The final contribution, Chapter 14, is a concluding essay by Paul Weirich. He stresses, among other things, the possibility of forming binding agreements as a way to handle the Prisoner's Dilemma. He asks us to reconsider a diachronic reformulation of the Prisoner's Dilemma in which two farmers sign a contract to help each other to harvest. One farmer first helps the other to harvest, after which the second helps the first. Without a binding agreement, this type of mutual aid would not be optimal for the players, at least not in all cases. Weirich then goes on to consider various ways in which one can compute the compensation the players should offer each other when agreeing to mutually aid each other.

1 Why all the fuss? The many aspects of the Prisoner's Dilemma

Ken Binmore

1.1 Immanuel Kant and my mother

I recall an old Dilnot cartoon in which Dilnot as a child is taken to task by his mother for some naughty behavior. Her rebuke takes the form of the rhetorical question:

> Suppose everybody behaved like that?

My own mother was fond of the same question. Like Dilnot in the cartoon, I would then silently rehearse the reasons that her logic was faulty. It is true that it would be bad if everybody were to behave asocially, but I am not everybody; I am just me. If my naughty behavior isn't going to affect anyone else, why does it matter to me what would happen if it did?

Benedict de Spinoza (1985) held the same view as my mother, as he reveals in the following passage on the irrationality of treachery:

> What if a man could save himself from the present danger of death by treachery? If reason should recommend that it would recommend it to all men.

Nor is he the only philosopher with a fearsome reputation for analytic rigor to take this line. Immanuel Kant (1993) elevates the argument into a principle of practical reasoning in his famous categorical imperative:

> Act only on the maxim that you would at the same time will to be a universal law.

Can such great minds really be wrong for the same reason that my mother was wrong?[1]

[1] Everybody agrees that it is *immoral* to behave asocially; the question is whether it is *irrational* to behave asocially.

It has become traditional to explain why the answer is *yes* using a simple game called the Prisoner's Dilemma. The analysis of this game is trivial and entirely unworthy of the attention that has been devoted to it in what has become an enormous literature. This chapter begins by defending the standard analysis, and continues by explaining why various attempts to refute it are fallacious.[2] However, the more interesting question is: Why all the fuss? How come some scholars feel such a deep need to deny the obvious?

I think the answer is to be found in the fact that there are circumstances in which an appeal to what everybody is doing is indeed a very good reason for doing the same thing yourself (Section 9). Otherwise the author of a display on an electronic board I passed when driving home on the freeway last night, which said:

> Bin your litter
> Everybody else does

would not have thought his argument compelling.

One can't exemplify the circumstances under which such arguments are valid using the Prisoner's Dilemma, but one can sometimes use the *indefinitely repeated* Prisoner's Dilemma for this purpose. This is a new and very different game in which the standard "one-shot" Prisoner's Dilemma is played repeatedly, with a small positive probability each time that the current repetition is the last. If all I succeed in doing in this chapter is to explain why it is important not to confuse the one-shot Prisoner's Dilemma with its indefinitely repeated cousin – or with any of the other games with which it has been confused in the past – I shall have achieved at least something.

1.2 Social dilemmas

This section looks at some examples that one might reasonably model using some version of the Prisoner's Dilemma. Psychologists refer to such situations as social dilemmas.

Some social dilemmas are mildly irritating facts of everyday social life. When waiting at an airport carousel for their luggage, everybody would be better off if everybody were to stand back so that we could all see our bags coming without straining our necks, but this isn't what happens. When waiting in a long line at a post office, everybody would be better off if

[2] The arguments mostly already appear in previous work (Binmore 1994, 2005, 2007b).

everybody were to conduct their business briskly when they get to the head of the line, but instead they join the clerks in operating in slow motion. Spectators at a football match would all be more comfortable staying in their seats when the game gets exciting, but they stand up instead and obscure the view of those behind.

As an example of a more serious social dilemma, consider what happens every year when we report our tax liability. If everybody employs a tax lawyer to minimize their liability, the government will have to raise the tax rate in order to maintain its revenue. Everybody who is not a lawyer will then be worse off because they now have to support a parasitic class of legal eagles without paying any less tax. Everybody would be better off if everybody were to dispense with a lawyer. Nevertheless, tax lawyers will get used.

Environmental versions of such social dilemmas are said to be Tragedies of the Commons, after a much quoted paper by Garret Hardin (1968), in which he drew attention to the manner in which common land eventually gets overgrazed when everybody is free to introduce as many animals as they like. The Tragedy of the Commons captures the logic of a whole spectrum of environmental disasters that we have brought upon ourselves. The Sahara Desert is relentlessly expanding southwards, partly because the pastoral peoples who live on its borders persistently overgraze its marginal grasslands. But the developed nations play the Tragedy of the Commons no less determinedly. We jam our roads with cars. We poison our rivers, and pollute the atmosphere. We fell the rainforests. We have plundered our fisheries until some fish stocks have reached a level from which they may never recover. As for getting together to make everybody better off by tackling the problem of global warming, we don't seem to be getting anywhere at all.

Congestion provides another arena for social dilemmas. For example, when driving from my home to my local airport, the signs all point one way but I go another. If everybody were to do what I do, my route would become hopelessly congested and the signs would be pointing in the correct direction for somebody in a hurry. However, everybody doesn't do what I do, and so the signs are lying – but who can blame whoever put them up? Even more difficult is the problem faced by officials who seek to persuade mothers to inoculate their children against a spectrum of possibly fatal childhood diseases. Because the inoculation itself is slightly risky, it isn't true – as the official propaganda insists – that a mother who wants the best for her child should necessarily have her child inoculated. What is best for her child depends on how many other children have been inoculated. In the case of measles, for example, a mother does best to inoculate her child only if more than 5% of other mothers have failed to do so.

Such stories invite commentary on moral issues. Is it moral for me to ignore the signs that say what the best way to the airport would be if everybody were to go that way? Is it moral for officials to tell mothers the lie that it is always best for them to have their children inoculated? But such moral issues are separate from the question that the Prisoner's Dilemma addresses, which is whether it can be in a rational person's self-interest to do what would be bad if everybody were to do it. Perhaps this is why Albert Tucker chose to illustrate the game with a story that marginalizes the moral issues when he first introduced the Prisoner's Dilemma to the world.[3]

1.3 Prisoner's Dilemma

Tucker's story is set in the Chicago of the 1920s. The District Attorney knows that Arturo and Benito are gangsters who are guilty of a major crime, but is unable to convict either unless one of them confesses. He orders their arrest, and separately offers each the following deal:

If you confess and your accomplice fails to confess, then you go free. If you fail to confess but your accomplice confesses, then you will be convicted and sentenced to the maximum term in jail. If you both confess, then you will both be convicted, but the maximum sentence will not be imposed. If neither confesses, you will both be framed on a minor tax evasion charge for which a conviction is certain.

In such problems, Arturo and Benito are the players in a game. Each player can choose one of two strategies called *hawk* and *dove*. The hawkish strategy is to betray your accomplice by confessing to the crime. The dovelike strategy is to stick by your accomplice by refusing to confess.

Game theorists assess what might happen to a player by assigning payoffs to each possible outcome of the game. The context in which the Prisoner's Dilemma is posed invites us to assume that neither player wants to spend more time in jail than they must. We therefore measure how a player feels about each outcome of the game by counting the number of years in jail he will have to serve. These penalties aren't given in the statement of the problem, but we can invent some appropriate numbers.

If Arturo holds out and Benito confesses, the strategy pair (*dove*, *hawk*) will be played. Arturo is found guilty, and receives the maximum penalty of ten years in jail. We record this result by making Arturo's payoff for (*dove*, *hawk*)

[3] The game was actually invented by the RAND scientists Dresher and Flood in 1950.

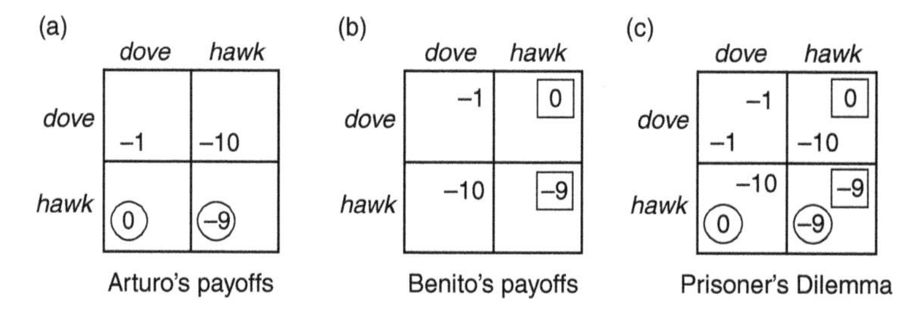

Figure 1.1 Prisoner's Dilemma. The payoffs fit Albert Tucker's original story in which two gangsters each seek to minimize how much time they will spend in jail. *Hawk* is a dominant strategy for each player, because it yields a better payoff than *dove* no matter what strategy the other player may choose.

equal to -10. If Benito holds out and Arturo confesses, (*hawk*, *dove*) is played. Arturo then goes free, and so his payoff for (*hawk*, *dove*) is 0. If Arturo and Benito both hold out, the outcome is (*dove*, *dove*). In this case, the District Attorney trumps up a tax evasion charge against both players, and they each go to jail for one year. Arturo's payoff for (*dove*, *dove*) is therefore -1. If Arturo and Benito both confess, the outcome is (*hawk*, *hawk*). Each player is found guilty, but since confession is a mitigating circumstance, each receives a penalty of only nine years. Arturo's payoff for (*hawk*, *hawk*) is therefore -9.

Figure 1.1a shows these payoffs for Arturo. The rows of this payoff table represent Arturo's two strategies. The columns represent Benito's two strategies. The circles show Arturo's best replies. For example, the payoff of –9 is circled in Figure 1.1a to show that *hawk* is Arturo's best reply to Benito's choice of *hawk* (because –9 is larger than –10). Figure 1.1b shows that Benito's preferences are the same as Arturo's, but here best replies are indicated by enclosing payoffs in a square. Figure 1.1c, which is the payoff table for a version of the Prisoner's Dilemma, shows both players' preferences at once.

Nash equilibrium. The fundamental notion in game theory is that of a Nash equilibrium.[4] A Nash equilibrium is any profile of strategies – one for each player – in which each player's strategy is a best reply to the strategies of the other players. The only cell of the payoff table of Figure 1.1c in which both payoffs are enclosed in a circle or a square is (*hawk*, *hawk*). It follows that (*hawk*, *hawk*) is the only Nash equilibrium of the Prisoner's Dilemma.[5]

[4] John Nash was the subject of the movie *A Beautiful Mind*, but the writers of the movie got the idea hopelessly wrong in the scene where they tried to explain how Nash equilibria work.

[5] Figure 1.1c only shows that (*hawk*, *hawk*) is the unique Nash equilibrium in *pure* strategies. In general, one also has to consider *mixed* strategies, in which players may randomize over their

Nash equilibria are of interest for two reasons. If it is possible to single out the rational solution of a game, it must be a Nash equilibrium. For example, if Arturo knows that Benito is rational, he would be stupid not to make the best reply to what he knows is Benito's rational choice. The second reason is even more important. An evolutionary process that always adjusts the players' strategy choices in the direction of increasing payoffs can only stop when it reaches a Nash equilibrium.

Paradox of rationality? Since (*hawk*, *hawk*) is the only Nash equilibrium of the Prisoner's Dilemma, game theorists maintain that it is the only candidate for the rational solution of the game. That is to say, rational players will both play *hawk*. The case for this conclusion is especially strong in the Prisoner's Dilemma because the strategy *hawk* strongly dominates the strategy *dove*. This means that each player gets a larger payoff from playing *hawk* than from playing *dove* – no matter what strategy the other player may choose.

The fact that rational players will both play *hawk* in the Prisoner's Dilemma is often said to be a *paradox of rationality* because both players would get a higher payoff if they both played *dove* instead. Both players will go to jail for nine years each if they both confess (by playing *hawk*), but both would go to jail for only one year each if they both refused to confess (by playing *dove*).

A whole generation of scholars set themselves the task of showing that game theory's resolution of this supposed "paradox of rationality" is mistaken – that rationality requires the play of *dove* in the Prisoner's Dilemma rather than *hawk*. Their reason for taking on this hopeless task is that they swallowed the line that this trivial game embodies the essence of the problem of human cooperation, and it would certainly be paradoxical if rational individuals were never able to cooperate! However, game theorists think it plain wrong to claim that the Prisoners' Dilemma embodies the essence of the problem of human cooperation. On the contrary, it represents a situation in which the dice are as loaded against the emergence of cooperation as they could possibly be. If the great game of life played by the human species were the Prisoners' Dilemma, we wouldn't have evolved as social animals! We therefore see no more need to solve some invented paradox of rationality than to explain why strong swimmers drown when thrown into Lake Michigan with their feet encased in concrete.

Much of the remainder of this chapter is devoted to explaining why no paradox of rationality exists. Rational players don't cooperate in the Prisoners'

set of pure strategies. They are mentioned here only so it can be observed that (*hawk*, *hawk*) remains the only Nash equilibrium even when mixed strategies are allowed.

Dilemma, because the conditions necessary for rational cooperation are absent in this game. The rest of the chapter reviews some of the circumstances under which it is rational to cooperate, and why it can then become sensible to pay very close attention to what everybody else is doing.

1.4 Public goods games

In public goods experiments, each of several subjects is assigned some money. The subjects can retain the money for themselves or contribute some or all of it to a public pool. The money contributed to the public pool is then increased by a substantial factor, and equal shares of the increased amount are redistributed to all the subjects – including those who contributed little or nothing.

In a simplified version of a public goods game, each of two players is given one dollar. They can either keep their dollar (*hawk*) or give it to the other player (*dove*) – in which case the dollar is doubled. The resulting game is shown in Figure 1.2a. It counts as a version of the Prisoner's Dilemma because *dove* is strictly dominated by *hawk*, but each player gets more if (*dove, dove*) is played than if (*hawk, hawk*) is played.

The immediate reason for drawing attention to the fact that the Prisoner's Dilemma can be seen as a kind of public goods game is that we can then appeal to surveys of a huge experimental literature on such games that has been independently surveyed by Ledyard (1995) and Sally (1995). The surveys show that the frequently repeated claim that experimental subjects cooperate (play *dove*) in laboratory experiments on the Prisoner's Dilemma is unduly

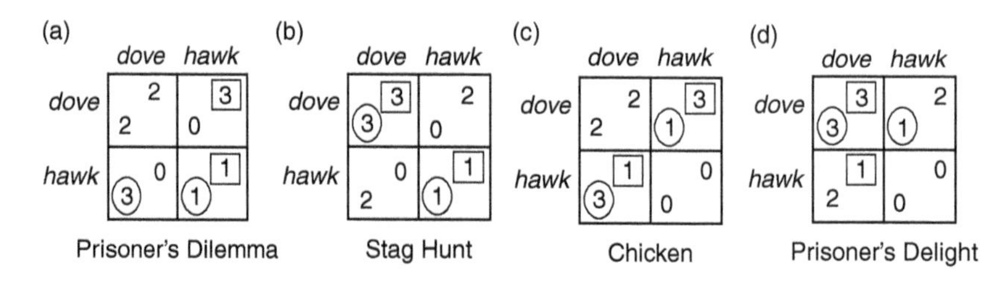

Figure 1.2 Other games. Figure 1.2a shows a version of the Prisoner's Dilemma as a public goods game as described in Section 1.4.

The payoffs are expressed in dollars rather than years in jail as in Figure 1.1. The same numbers are used in Section 1.5 when explaining how the Prisoner's Dilemma is obtained using the theory of revealed preference to avoid the need to assume that the players seek to maximize or minimize some particular quantity like dollars or years in jail. The payoffs are then measured in units of utility called utils. The remaining figures show some standard games that result when the players do *not* reveal preferences that accord with the assumptions built into the Prisoner's Dilemma.

optimistic. Slightly more than half of inexperienced subjects play *dove*, and the rest play *hawk* (Camerer 2003). However, after playing the game ten times against a new opponent each time, 90 percent or more of subjects end up playing *hawk* – a result that accords with the game theory prediction that trial-and-error adjustment in the direction of higher payoffs should be expected to converge on a Nash equilibrium.

1.5 Revealed preference

Controversy over the Prisoner's Dilemma persists largely because critics of the orthodox analysis focus not on the game itself, but on the various stories used to introduce the game. They then look for a way to retell the story that makes it rational to cooperate by playing *dove*. If successful, the new story necessarily leads to a new game in which it is indeed a Nash equilibrium for both players to choose *dove*. One can, for example, easily create versions of the Prisoner's Delight of Figure 1.2d by assuming that Benito and Arturo care for each other's welfare (Binmore 2007b). But showing that it is rational to cooperate in some game related to the Prisoner's Dilemma should not be confused with the impossible task of showing that it is rational to cooperate in the Prisoner's Dilemma itself. To press home this point, the current section continues with an account of what needs to be true of a story for it to be represented by the Prisoner's Dilemma.

The players will now be called Adam and Eve to emphasize that we have abandoned the story of gangsters in Chicago, but how do we come up with a payoff table without such a story? The orthodox answer in neoclassical economics is that we discover the players' preferences by observing the choices they make (or would make) when solving one-person decision problems.

Writing a larger payoff for Adam in the bottom-left cell of the payoff table of the Prisoners' Dilemma than in the top-left cell therefore means that Adam would choose *hawk* in the one-person decision problem that he would face if he knew in advance that Eve had chosen *dove*. Similarly, writing a larger payoff in the bottom-right cell means that Adam would choose hawk when faced with the one-person decision problem in which he knew in advance that Eve had chosen *hawk*.

The very definition of the game therefore says that *hawk* is Adam's best reply when he knows that Eve's choice is *dove*, and also when he knows her choice is *hawk*. So he doesn't need to know anything about Eve's actual choice to know his best reply to it. It is rational for him to play *hawk* whatever strategy she is planning to choose. Nobody ever denies this utterly trivial argument. Instead, one is told that it can't be relevant to anything real because it reduces the analysis of the Prisoners' Dilemma to a tautology. But who would similarly say that the tautology 2+2=4 is irrelevant to anything real?

In summary, to obtain a version of the Prisoner's Dilemma, we need that Adam would choose (*hawk, dove*) over (*dove, dove*), and (*hawk, hawk*) over (*dove, hawk*) if he had the opportunity to choose. For the appearance of a paradox of rationality, we also need that Adam would choose (*dove, dove*) over (*hawk, hawk*). Corresponding assumptions must also be made for Eve.

To respond that Adam wouldn't choose (*hawk, dove*) over (*dove, dove*) is rather like denying that 2+2=4 because the problem should really be to compute 2+3 instead of 2+2. This is not to deny that there may be many reasons why a real-life Adam might choose (*dove, dove*) over (*hawk, dove*) if offered the choice, He might be in love with Eve. He might get pleasure from reciprocating good behavior. He might have agreed with Eve before the game that they will both play *dove* and hates breaking promises.

Such tales may be multiplied indefinitely, but they are irrelevant to an analysis of the Prisoner's Dilemma. If we build into our game the assumption that Adam would choose (*dove, dove*) over (*hawk, dove*) and that Eve would similarly choose (*dove, dove*) over (*dove, hawk*), we end up with a version of the Stag Hunt Game[6] of Figure 1.2b (provided that everything else stays the same). It is true that it is a Nash equilibrium for both Adam and Eve to play *dove* in the Stag Hunt Game, but it doesn't follow that it is rational to play *dove* in the Prisoner's Dilemma, because the Prisoner's Dilemma is not the Stag Hunt Game.

Figure 1.2 shows all the symmetric games that can be generated by altering the preferences attributed to the players without introducing new payoff numbers. Chicken does not even have a symmetric Nash equilibrium in pure strategies.

1.6 Squaring the circle?

This section reviews the fallacious arguments of the philosophers Nozick (1969) and Gauthier (1986) that purport to show that it is rational to cooperate in the Prisoner's Dilemma. Other fallacies, notably the Symmetry Fallacy of Rapoport (1966) and Hofstadter (1983), are also reviewed in my book *Playing Fair* (Binmore 1994).

[6] The game is said to instantiate Jean-Jacques Rousseau's story of a stag hunt, but my reading of the relevant passages makes him just another philosopher telling us that it is rational to cooperate in the Prisoner's Dilemma (Binmore 1994).

1.6.1 The fallacy of the twins

Two rational people facing the same problem will come to the same conclusion. Arturo should therefore proceed on the assumption that Benito will make the same choice as him. They will therefore either both go to jail for nine years, or they will both go to jail for one year. Since the latter is preferable, Arturo should choose *dove*. Since Benito is his twin, he will reason in the same way and choose *dove* as well.

This fallacious argument is attractive because there are situations in which it would be correct. For example, it would be correct if Benito were Arturo's reflection in a mirror, or if Arturo and Benito were genetically identical twins and we were talking about what genetically determined behavior best promotes biological fitness (see below). The argument would then be correct because the relevant game would no longer be the Prisoner's Dilemma. It would be a game with only one player. But the Prisoner's Dilemma is a two-player game in which Arturo and Benito choose their strategies *independently*.

Where the twins fallacy goes wrong is in assuming that Benito will make the same choice in the Prisoner's Dilemma as Arturo *whatever* strategy he might choose. This can't be right, because one of Arturo's two possible choices is irrational. But Benito is an independent rational agent. He will behave rationally whatever Arturo may do.

The twins fallacy is correct only to the extent that rational reasoning will indeed lead Benito to make the same strategy choice as Arturo in the Prisoner's Dilemma (if Arturo also chooses rationally). Game theorists argue that this choice will be *hawk* because *hawk* strongly dominates *dove*.

Hamilton's rule. Suppose Arturo and Benito do not choose independently. They will not then be playing the Prisoner's Dilemma, but they will be playing a game that biologists think is important in explaining the evolution of altruism.

An important case was first considered by Bill Hamilton.[7] Suppose that Adam and Eve are relatives whose behavior is determined by a single gene that is the same in each player with probability r.[8] Both players can act altruistically (play *dove*) by conferring a benefit b on the other at cost of

[7] In biology, the payoffs in a game are interpreted as (inclusive) fitnesses, expressed as the number of copies of the operant gene transmitted to the next generation.

[8] The probability r is the *degree of relationship* between Adam and Eve. For a recently mutated gene, $r = 1$ for identical twins, $r = 1/2$ for siblings, and $r = 1/8$ for full cousins. Hamilton's analysis has been extended by himself and others to cases in which correlations arise for reasons other than kinship.

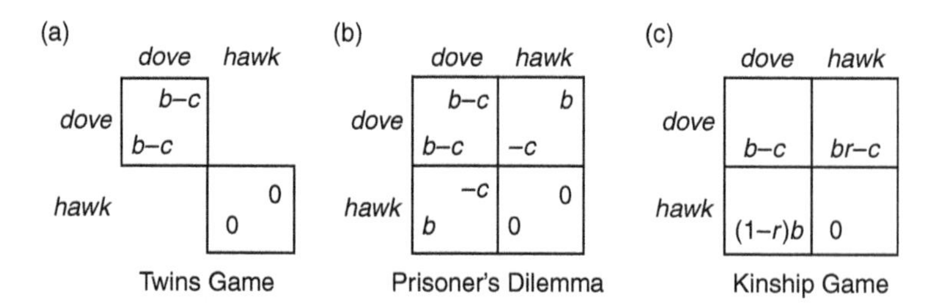

Figure 1.3 Hamilton's rule. The Kinship Game begins with a chance move that selects the Twins Game with probability r, and the Prisoner's Dilemma with probability $1 - r$. Figure 1.3c shows Adams's (average) payoffs in this game, where the columns correspond to the strategies that Eve might play when she doesn't share the operant gene with Adam.

c (where $b > c > 0$); or they can act selfishly (play *hawk*) by doing nothing. Hamilton's rule says that altruism will be selected for if

$$br > c$$

This inequality says that the benefit to your relative weighted by your degree of relationship must exceed the cost of providing the benefit. How come?

The game Adam and Eve's *genes* are playing opens with a chance move that selects the one-player game of Figure 1.3a with probability r and the version of the Prisoner's Dilemma of Figure 1.3b with probability $1-r$. Figure 1.3c shows Adam's payoffs averaged over these two possibilities.[9] We need to write down two cases for each of Adam's strategies to take account of the two possibilities that can occur when Eve's gene is not the same as Adam's. In the latter case, the alien gene to which Eve then plays host may specify either *dove* or *hawk* independently of Adam's strategy.

Having written down Adam's payoff matrix, it only remains to observe that *dove* now strongly dominates *hawk* if and only if Hamilton's rule holds. It is then rational to play *dove* in the Kinship Game we have been studying, but it remains irrational to do so in the Prisoner's Dilemma.

1.6.2 The transparent disposition fallacy

This fallacy depends on two doubtful claims. The first is that rational people have the willpower to commit themselves in advance to playing games in a

[9] For example, the top-right payoff is calculated as $br - c = (1 - r)(-c) + r(b - c)$, where the final term registers what Adam expects to get from playing himself when he chooses *dove*.

particular way. The second is that other people can read our body language well enough to know when we are telling the truth. If we truthfully claim that we have made a commitment, we will therefore be believed. If these claims were correct, our world would certainly be very different! Rationality would be a defense against drug addiction. Poker would be impossible to play. Actors would be out of a job. Politicians would be incorruptible. However, the logic of game theory would still apply.

As an example, consider two possible mental dispositions called *clint* and *john*. The former is named after the character played by Clint Eastwood in the spaghetti westerns. The latter commemorates a hilarious movie I once saw in which John Wayne played the part of Genghis Khan. To choose the disposition *john* is to advertise that you have committed yourself to play *hawk* in the Prisoners' Dilemma no matter what. To choose the disposition *clint* is to advertise that you are committed to play *dove* in the Prisoners' Dilemma if and only if your opponent is advertising the same commitment. Otherwise you play *hawk*.

If Arturo and Benito can commit themselves transparently to one of these two dispositions before playing the Prisoners' Dilemma, what should they do? Their problem is a game in which each player has two strategies, *clint* and *john*. The outcome of this Film Star Game will be (*hawk*, *hawk*) in the ensuing Prisoner's Dilemma unless both players choose *clint*, in which case it will be (*dove*, *dove*). It is obviously a Nash equilibrium for both players to choose *clint* in the Film Star Game, with the result that both will be committed to play *dove* in the Prisoner's Dilemma.

Advocates of the transparent disposition fallacy think that this result shows that cooperation is rational in the Prisoners' Dilemma. It would be nice if they were right in thinking that real-life games are really all Film Star Games of some kind – especially if one could choose to be Adam Smith or Charles Darwin rather than John Wayne or Clint Eastwood. But even then they wouldn't have shown that it is rational to cooperate in the Prisoners' Dilemma. Their argument only shows that it is rational to play *clint* in the Film Star Game.

1.7 Reciprocal altruism

The fact that rational players will not cooperate in the Prisoner's Dilemma does not imply that rational players will never cooperate. They will cooperate when they play games in which it is rational to cooperate. For example, cooperation is rational in the Prisoner's Delight of Figure 1.2, versions of which are easily generated by assuming that the players actively care about each other's welfare – as with relatives or lovers (Binmore 2006).[10]

The Stag Hunt Game is more interesting because it provides an example of a game with multiple equilibria. If a rational society found itself operating the equilibrium in which everybody plays *hawk*, why wouldn't everybody agree to move to the equilibrium that everybody prefers in which everybody plays *dove*? If so, shouldn't we reject the *hawk* equilibrium as irrational? To see why most (but not all) game theorists reject the idea that rationality implies anything more than that an equilibrium will be played,[11] consider the following argument that shows why an agreement to shift from the *hawk* equilibrium to the *dove* equilibrium can't be sustained on purely rational grounds.

Suppose that Adam and Eve's current social contract in the Stag Hunt is the Nash equilibrium in which they both play *hawk*. However hard Adam seeks to persuade Eve that he plans to play *dove* in the future and so she should follow suit, she will remain unconvinced. The reason is that whatever Adam is actually planning to play, it is in his interests to persuade Eve to play *dove*. If he succeeds, he will get 3 rather than 0 if he is planning to play *dove*, and 2 rather than 1 if he is planning to play *hawk*. Rationality alone therefore doesn't allow Eve to deduce anything about his plan of action from what he says, because he is going to say the same thing no matter what his real plan may be! Adam may actually think that Eve is unlikely to be persuaded to switch from *hawk* and hence be planning to play *hawk* himself, yet still try to persuade her to play *dove*.

This Machiavellian story shows that attributing rationality to the players isn't enough to resolve the equilibrium selection problem in games with multiple equilibria – even in a seemingly transparent case like the Stag Hunt. If Adam and Eve continue to play *hawk* in the Stag Hunt, they will regret their failure to coordinate on playing *dove*, but neither can be accused of being irrational, because both are doing as well as they can given the behavior of their opponent.

[10] David Hume famously said that there would be nothing *irrational* in his preferring the destruction of the whole world to scratching his finger. Rationality in game theory similarly implies nothing whatever about the preferences of the players. It is understood to be about the *means* players employ in seeking to achieve whatever (consistent) set of *ends* they happen to bring to the table.

[11] Some game theorists argue that Nash equilibrium should be refined to something more sophisticated, but these arguments do not apply in the Stag Hunt Game.

The standard response is to ask why game theorists insist that it is irrational for people to trust one another. Wouldn't Adam and Eve both be better off if both had more faith in each other's honesty? But nobody denies that Adam and Eve would be better off if they trusted each other, any more than anybody denies that they would be better off in the Prisoner's Dilemma if they were assigned new payoffs that made them care more about the welfare of their opponent. Nor do game theorists say that it is irrational for people to trust each other. They only say that it isn't rational to trust people without a good reason: that trust can't be taken upon trust.

To understand when it is rational to trust other people without begging the question by building a liking for trustworthy behavior or altruism into their preferences,[12] game theorists think it necessary to turn to the study of repeated games. The reason is that rational reciprocity can't work unless people interact repeatedly without a definite end to their relationship in sight. If the reason I scratch your back today is that I expect you will then scratch my back tomorrow, then our cooperative arrangement will unravel if we know that there will eventually be no tomorrow. As David Hume put it:

> I learn to do service to another, without bearing him any real kindness, because I foresee, that he will return my service in expectation of another of the same kind, and in order to maintain the same correspondence of good offices with me and others. And accordingly, after I have serv'd him and he is in possession of the advantage arising from my action, he is induc'd to perform his part, as foreseeing the consequences of his refusal. (Hume 1739: III.V)

Repeated games. Game theorists rediscovered David Hume's insight that reciprocity is the mainspring of human sociality in the early fifties when characterizing the outcomes that can be supported as equilibria in a repeated game. The result is known as the *folk theorem*, since it was discovered independently by several game theorists in the early 1950s (Aumann and Maschler 1995). The theorem tells us that rational people do not need to care for each other in order to cooperate. Nor is any coercion needed. It is only

[12] Some behavioral economists argue that their experiments show that people do indeed have a liking for socially desirable traits like trustworthiness built into their natures – that we are all Dr. Jekylls to some extent. I am doubtful about the more extravagant of such claims (Binmore and Shaked 2010). But the folk theorem of repeated game theory shows that even a society of Mr. Hydes is capable of high levels of rational cooperation.

necessary that the players be sufficiently patient, and that they know they are to interact together for the foreseeable future. The rest can be left to their enlightened self-interest, provided that they can all monitor each other's behavior without too much effort – as was the case when we were all members of small hunter-gatherer communities.

To see why, we need to ask what outcomes can be sustained as Nash equilibria when a one-shot game like the Prisoner's Dilemma or any of the other games of Figure 1.2 is repeated indefinitely often. The answer provided by the folk theorem is very reassuring. Any outcome whatever of the one-shot game – including all the outcomes that aren't Nash equilibria of the one-shot game – can be sustained as Nash equilibrium outcomes of the *repeated* game, provided that they award each player a payoff that is larger than the player's minimax payoff in the one-shot game.

The idea of the proof is absurdly simple.[13] We first determine how the players have to cooperate to obtain a particular outcome. For example, in the repeated Prisoners' Dilemma, Adam and Eve need only play *dove* at each repetition to obtain an average payoff of 2. To make such cooperative behavior into a Nash equilibrium, it is necessary that a player who deviates be punished sufficiently to make the deviation unprofitable. This is where the players' minimax payoffs enter the picture, because the worst punishment that Eve can inflict on Adam in the long run is to hold him to his minimax payoff.

In the Prisoner's Dilemma, the minimax payoff for each player is 1, because the worst that one player can do to the other is play *hawk*, in which case the victim does best to respond by playing *hawk* as well. The folk theorem therefore tells us that we can sustain the outcome in which both players always play *dove* as a Nash equilibrium in the indefinitely repeated game.

The kind of equilibrium strategy described above is called the *grim* strategy,[14] because the punishment that keeps potential deviants on the strait and narrow path is the worst possible punishment indefinitely prolonged.

[13] Provided that the probability that any particular repetition of the one-shot game is the last is sufficiently small.

[14] The importance of reciprocity in sustaining Nash equilibria in repeated games has been confused by the exaggerated claims made by Axelrod (1984) for the particular strategy *Tit-for-Tat*. It is true that (*Tit-for-Tat, Tit-for-Tat*) is a Nash equilibrium for the indefinitely repeated Prisoner's Dilemma if the players care enough about the future, but the same can be said of an infinite number of other strategy profiles, notably (*grim, grim*). The other virtues claimed for *Tit-for-Tat* are illusory (Binmore 2001).

One sometimes sees this strategy in use in commercial contexts where maintaining trust is vital to the operation of a market. To quote a trader in the New York antique market:

> Sure I trust him. You know the ones to trust in this business. The ones who betray you, bye-bye. *New York Times*, August 29, 1991.

However, one seldom sees such draconian punishments in social life. Indeed, most of the punishments necessary to sustain the equilibria of ordinary life are administered without either the punishers or the victim being aware that a punishment has taken place. Shoulders are turned slightly away. Eyes wander elsewhere. Greetings are imperceptibly gruffer. These are all warnings that you need to fall in line lest grimmer punishment follow. I recall being particularly delighted when I discovered anthropological accounts of higher stages of punishment observed among hunter-gatherer societies, because they mirror so accurately similar phenomena that the academic world uses to keep rogue thinkers in line. First there is laughter. If this doesn't work – and who likes being laughed at – the next stage is boycotting. Nobody talks to the offending party, or refers to their research. Only the final stage is maximal: persistent offenders are expelled from the group, or are unable to get their work published.

Criticisms. Three criticisms of applications of the folk theorem are common. The first denies its assumption that any deviations from equilibrium will necessarily be observed by the other players. This is probably not a bad assumption in the case of the small bands of hunter-gatherers in which we developed as social animals, or in the small towns of today where everybody knows everybody else's business. But what of modern city life, where it isn't possible to detect and punish deviants often enough to deter cheating? The reply is simple. The reason we spend so much time locking our doors and counting our change is that we know that we are always at risk of being victimized when out and about in the big city. That is to say, it is true that traditional folk theorems do not apply in a big city, and that is why we see more bad behavior there. The criticism is therefore valid insofar as it denies that folk theorems show that cooperation is always rational in all societies, but nobody argues otherwise.

The second criticism denies that it is necessarily rational to punish if someone should deviate from equilibrium play. This is a good point, but the folk theorem also holds with Nash equilibria replaced by subgame-perfect equilibria. The strategies specified by such an equilibrium demand that the players plan to play a Nash equilibrium not only in the whole game but in all its subgames – whether the subgame would be reached or not in the original equilibrium. Players required to punish a deviant who forces play into a subgame that would not otherwise be reached can then be incentified to carry out the punishment by specifying an equilibrium for the subgame in which the specified punishers would themselves be punished if they failed to carry out the punishment. In brief, game theory has an answer to the perennial question: Who guards the guardians? They can be made to guard each other.

The third criticism denies that people are as selfish as the folk theorem assumes. But the folk theorem doesn't assume that people are selfish. It doesn't assume anything at all about what people want or don't want. It is true that they are assumed to maximize their own utility functions, but who is to say that their utility functions are selfish rather than incorporating a deep and abiding concern for the welfare of others? Economists admittedly seldom think that such models of human nature are very realistic, but that is another matter.

1.8 Social norms

The folk theorem not only gives conditions under which people can cooperate rationally, it shows that there are generally an infinite number of ways in which people can cooperate. In solving one problem, we therefore create another. How do we decide which of all the cooperative equilibria is to be regarded as the solution to the game?

The repeated Prisoner's Dilemma is misleading in this respect because it is symmetric (looks exactly the same to both players). In such games one is offered an obvious solution to the equilibrium selection problem: choose an equilibrium in which both players get the same payoff. But what of the asymmetric coordination games of which our social life largely consists?

Tom Schelling (1960) followed David Hume (1978) in arguing that social norms or conventions exist for solving the equilibrium selection problem in such games. In my own work, I focus on the evolution of fairness norms for this purpose (Binmore 2005). However, the immediate point is that rationality has no more of a role to play in determining which social norm will evolve to select an equilibrium to serve as the solution of a coordination game than it has in determining whether we drive on the left or the right in the Driving Game. So how do we figure out what equilibrium we ought to play when confronted with an unfamiliar game, or a familiar game being played in an unfamiliar context?

The answer returns us to the puzzle of why it is commonly thought that it is rational to do whatever everybody else is doing. It is indeed rational to follow this maxim when in doubt of what to do in a game with multiple equilibria. To know what is best to do, I need to know or guess what everybody else is doing. If they are acting on the assumption that everybody will play a particular equilibrium, then it is optimal for me to play my part in operating that equilibrium too. For example, if everybody else is driving on the left, it would be stupid for me to do anything other than drive on the left as well.

I think all the fuss about the Prisoner's Dilemma arises from a failure to understand the limitations of this principle. It isn't a substitute for the kind of reasoning that leads us to the notion of a Nash equilibrium. It is a recognition of the fact that rationality alone can't tell us what equilibrium to play in a game with multiple equilibria.

1.9 Conclusions

When is it rational to do what everybody else is doing? If you like being with other people doing what they do – as is presumably the case with people who attend discos – then the question answers itself. But what if you don't like doing what everybody else is doing just for its own sake? It can then still make sense to do what everybody else is doing if you are all playing one of the many coordination games of which human social life mostly consists.

Coordination games[15] commonly have many Nash equilibria. If there is common knowledge of rationality or if social evolution has had long enough to act, then one of these equilibria will be played. But rational or evolutionary considerations can't tell you in advance *which* equilibrium will be played without some information about the intentions of the other players. This information becomes abundantly available if you learn that amongst the group of people from which the players of the game are drawn, everybody has always coordinated on one particular equilibrium in the past. With this information, it becomes best for you to play your part in operating that particular equilibrium. If one knows, for example, that everybody in Japan coordinates on the equilibrium in the Driving Game in which all players drive on the left, then you will want to drive on the left as well when in Japan.

[15] Of which the indefinitely repeated Prisoner's Dilemma is a leading example.

However, this conclusion only applies if the principle of doing what everybody else does is restricted to choosing between rival *equilibria*. If rationality or evolution has not succeeded in coordinating the behavior of other people on some particular equilibrium of whatever game is being played – as in certain countries where I have played the Driving Game – then it will no longer necessarily be best for you to do whatever everybody else is doing, because there is then no guarantee that such behavior is a best response to the behavior of the other players in the game.

The same reasoning applies to *not* doing what everybody else is *not* doing. But this is a long way from the categorical imperative that tells us *not* to do what would be bad if everybody were to do it. The orthodox game-theoretic analysis of the Prisoner's Dilemma shows that this principle can't be defended purely by appeals to rationality as Kant claims. However, I have argued elsewhere that a suitably refined version of the principle is incorporated in the fairness norms that societies often use to solve equilibrium selection problems in new situations (Binmore 2005). That is to say, the categorical imperative has a role in discussions of what is moral, but not in what is rational.

2 How I learned to stop worrying and love the Prisoner's Dilemma

David Gauthier

2.1 Structures of interaction and theories of practical rationality

Two persons, each with but two possible actions. Four possible outcomes. The Prisoner's Dilemma is the simplest structure of interaction in which, if each person, in choosing that action which, given the other person's choice, maximizes his own expected utility, the outcome will afford each of the persons less (expected) utility than would the outcome were each to choose his other, non-maximizing action.

In many situations with this structure, it is plausible to think of a person who chooses not to maximize as cooperating, or seeking to cooperate, with the other person. They do better working together than working alone. The person who seeks to maximize his utility, given the other's choice, may then be represented, perhaps somewhat pejoratively, as defecting from the cooperative arrangement. But since Ken Binmore, who is one of the leading defenders of the rationality of maximizing choice, is happy to use the terminology of cooperate and defect, I shall follow his example.[1]

Most economists, like Binmore, and some philosophers, believe that it is rational for persons to choose the action that they judge will maximize their expected utility. In the Prisoner's Dilemma, each person sees that, if the other cooperates, then he will maximize his expected utility by defecting. Each also sees that if the other defects, he will also maximize his expected utility by defecting. So defecting is the maximizing choice for him, whatever the other does. So Binmore claims – and it is here that I shall part company with him – it is rational for each to defect. Although each would expect greater utility if they cooperated.

In situations more complex than the Prisoner's Dilemma, there may be no choice among a person's possible actions that will maximize his expected utility whatever the other persons do. But suppose we restrict ourselves to

[1] Binmore (1991: 133).

interactions with finitely many persons each with finitely many possible actions – hardly onerous restrictions. And suppose each person is able to randomize at will over his actions. Then, as John Nash has proved,[2] there must be at least one set of choices (including randomized choices), one for each person such that each member of the set maximizes the chooser's utility provided that every person chooses his action from the set. The outcome will then be in equilibrium, since no person can increase his expected utility by choosing a different action.

Suppose that our theory of practical rationality prescribes a choice for each person involved in an interaction. Each choice, it will be argued, must maximize the person's utility, given the choices prescribed for the others. For if it did not, then some person would be able to increase his utility by choosing contrary to what is prescribed by the theory. And the theory would then fail to satisfy the basic assumption that it is rational for persons to maximize their expected utility. Since this assumption is at the heart of rational choice theory, questioning it may seem on a par with tilting at windmills. But question it I shall, with the Prisoner's Dilemma at the core of my argument. And since I want to appeal to the Dilemma, I need to defend what I am doing against the charge that I am changing the subject, and discussing some other structure of interaction.[3]

2.2 Revealed preference and rational choice

Consider these two statements, taken from D.M. Winch's *Analytical Welfare Economics* and typical of the views held by economists: (1) "...we assume that individuals behave rationally and endeavor to maximize utility";[4] (2) "We assume that individuals attempt to maximize utility, and define utility as that which the individual attempts to maximize."[5] The first of these statements identifies rational behavior with utility-maximizing behavior. The second defines utility as what a person seeks to maximize. Neither is correct – or so I shall argue.

The second is at the core of revealed preference theory. On this view a person's preferences, which are measured by utility, are revealed in (and only in) his choices. We should not think of utilities as inputs with choices as outputs, but rather we should take a person's choices as determining his utilities. Whatever his choices maximize is his utility.

[2] Nash (1951). [3] As Binmore claims (1991: 140). [4] Winch (1971: 21).
[5] Winch (1971: 25).

This has rather dire consequences for the Prisoner's Dilemma. For if revealed preference theory is sound, then if one or both of the persons were to choose to cooperate, it follows that for him or them, cooperation would have greater expected utility than defection, contrary to what is assumed in the Dilemma. Indeed, the Prisoner's Dilemma would metamorphose into some other structure of interaction. So in purporting to defend cooperation, I would actually be arguing about some other game. Binmore would be right.

But my objection to revealed preference theory is not only that it prevents us from making good sense of the Prisoner's Dilemma. Revealed preference theory is a conceptual straitjacket. It treats two distinct phenomena, evaluation and choice, as if they were one. Counter-preferential choice cannot even be formulated in a theory that treats utilities as determined by choices. If utilities are revealed by whatever is chosen, then choice cannot run afoul of preferences.

Among the capacities of rational agents is the ability to evaluate the possible outcomes of their actions as better or worse from their own perspective. Preference is linked to evaluation; an agent prefers one outcome to another just in case it is, from his point of view, better than the other. The possible outcomes can be ranked from best to worst and, given appropriate idealizing conditions, this ranking determines a real-valued function that assigns a numerical utility to each outcome. The idealizing conditions are of no concern to my present argument, so I shall put them aside.

The ability to evaluate is of course manifested far more widely than directly in the context of choice. Aesthetic evaluation is a reaction to certain experiences that in themselves need have no bearing on what we might do. Ranking Picasso as a greater painter than Georges Braque may be relevant if we should have to choose between seeing an exhibition of Picasso's works and seeing an exhibition of Braque's (though it need not be relevant), but the point of ranking Picasso and Braque is not likely to be to guide such a choice.

I want therefore to suggest that we treat a utility function as simply an instance of an evaluative ranking, where such a ranking is in itself quite independent of choice. To be sure, it would be irrational not to recognize the bearing of a ranking of possible outcomes of one's actions on one's choice of which outcome to endeavor to realize. But it does not follow from this that the ranking provides the sole ground for rational choice, much less that one's choices must reveal the rankings that make up one's utility function. That rational choice reveals a person's utilities is a substantive thesis about the relation between choice and preference and, I shall argue, a mistaken one.

2.3 Pareto-optimality

I should at this point emphasize that I am not accusing game theorists and economists of any error in their formal analysis. They are exploring interaction among agents whose choices can be represented as maximizing. Such agents will seek to act to obtain their greatest expected utility given the actions of the other agents, who are represented similarly. If the agents' expectations of their fellows' choices are all correct, then each agent will select his best reply to the others, and the outcome will be "an *equilibrium* of some kind."[6] The real world interest in this analysis will depend on being able to treat real agents as game-theoretic maximizers. But I shall argue that the Prisoner's Dilemma shows that rationality should not be identified with expected utility maximization. And this brings us back to the first of Winch's statements that I reject – "we assume that individuals behave rationally and endeavor to maximize utility." To identify rational behavior with utility maximization is to deny the possibility of non-maximizing rationality. We should understand this identification as a substantive thesis, and one that I shall reject. Of course in many situations it is true that rational behavior will be utility-maximizing. But a brief consideration of best-reply reasoning in the Prisoner's Dilemma and other structures of interaction should convince us of its inadequacy.

There are two claimants to the demands of rational behavior in the Prisoner's Dilemma. The first claimant, supported by the economists and their allies, is the familiar one that says that each agent acts rationally in seeking to maximize his expectation of utility, which leads, as we have seen, to the rational outcome being a mutual best reply. Since in the Prisoner's Dilemma each person does better to defect rather than cooperate whatever the other does, the rational outcome is mutual defection.

But this seems absurd. The rational outcome, it is being claimed, is for each person third best of the four possible outcomes, but they could achieve the second best for each, if they were both to choose to cooperate. Let us then introduce the second claimant, which says that each acts rationally in seeking an outcome that maximizes his expectation of utility, given the utilities it affords the others. We can make this more precise.

We call one outcome Pareto-superior to another, just in case it is considered better by everyone than the other. And we call an outcome Pareto-optimal, just in case there is no outcome Pareto-superior to it. In the Prisoner's Dilemma the

[6] Binmore (1991: 140).

outcome of mutual cooperation is Pareto-superior to the outcome of mutual defection, and is Pareto-optimal. A Pareto-optimal outcome must afford each person a utility that is maximal given the utilities it affords the others. Instead of looking for the rational outcome among the mutual best replies (in the Prisoner's Dilemma the only mutual best reply is joint defection), one should look to the Pareto-optimal outcomes, and in the Prisoner's Dilemma the symmetry of the situation ensures that joint cooperation is selected.

Appeal to a Pareto condition similar to what I am proposing has been introduced by Ned McClennen[7]. But he introduces it in the very limited surrounds of pure coordination. And I am not persuaded either that rational choice in such situations is addressed by a Pareto-condition, or that we can extrapolate usefully from what would be rational in pure coordination to what is rational in other contexts. In pure coordination situations, the persons are in full agreement on the valuation of the possible outcomes, so that we need only a single number to represent the utility of each outcome. We may then call an outcome *optimal* with no qualifier, just in case no other outcome affords greater utility. We can say that the outcome is Pareto-optimal, and in equilibrium, but its straightforward optimality is what recommends it. McClennen does not argue in this way, but supposes that Pareto-optimality is the appropriate rationality condition for pure coordination. He then extends his account to interactions that involve both coordination and competition, and tries to find a role for the Pareto-condition in this broader area of rational choice. But if the optimality condition relevant to pure coordination is not Pareto-optimality then McClennen's extrapolation would seem to lack grounding.

In my view, Pareto-optimality comes into play when there is no straightforwardly optimal outcome, and also, of course, when not all outcomes are Pareto-optimal, as they are in zero-sum situations. And it stands as a contrast to best reply or equilibrium considerations. These address the acts each person performs in relation to the actions of the others. It addresses instead the utility each person obtains in relation to the utilities of the others. In interactions, irrationality lies not in the failure of individuals to make the most of their own particular situation, but rather in the failure of all those party to the interaction to make the most of their joint situation. If the outcome is Pareto-optimal, then they have made the most, in the sense that no one has forgone a benefit that he could have enjoyed without a cost to

[7] McClennen (2008, 36–65, esp. 44–51). When I read his essay, I find extensive passages that I could have written myself in stating my own view.

some other person. Pareto-optimality ensures that the persons have interacted efficiently. And while this may not always be sufficient for rational interaction, it is necessary – or so I claim.

A person acts rationally, on this view, in seeking to bring about an outcome which is Pareto-optimal. (I shall henceforth abbreviate this to "P-optimal.") But this need not always be possible. The behavior of the other parties in an interaction may preclude a person from acting to bring about any P-optimal outcome (or any acceptable P-optimal outcome, where acceptability has yet to be explicated). The Prisoner's Dilemma will illustrate this last case. If Column will defect, then Row clearly should also defect. But if Row were to cooperate, then the outcome (0, 4) would be P-optimal. (0, 4) satisfies the definition of P-optimality, since any alternative must be worse for Column. But it is clearly unacceptable for Row. P-optimality cannot be a sufficient condition for rationality.

2.4 Best-reply theories and P-optimal theories

The Prisoner's Dilemma provides a clear representation of an interaction in which there is a sharp contrast between the actions required and the resulting outcomes afforded by the two views of rational behavior that I have distinguished. But the Prisoner's Dilemma does not reveal all that either view must address. In the Prisoner's Dilemma there is only one outcome that satisfies the mutual best reply (or equilibrium) condition. Many interactions, however, possess multiple equilibria, and a fully developed theory of rational choice would provide a way to choose among them. Furthermore, in the Prisoner's Dilemma there is only one outcome that satisfies both the P-optimal condition, and an obvious symmetry condition. But most interactions are not strongly symmetrical and there are multiple P-optimal outcomes. A fully developed theory of rational choice would provide a way to choose among them.

I shall speak of theories of practical rationality that satisfy the best reply or equilibrium condition as *best-reply* theories, and those that satisfy the P-optimal condition as *P-optimal* theories. It is evident that in the Prisoner's Dilemma, the P-optimal outcome is P-superior to the best-reply outcome. One might think that this could be generalized. One might suppose that if in some interaction no best-reply outcome is P-optimal, then the outcome given by a P-optimal theory must be P-superior to the outcome given by a best-reply theory.

We must proceed carefully here. If an equilibrium outcome is not P-optimal, then of course there must be some possible outcomes that are

P-superior to it. But there need be no one outcome that is P-superior to every non-P-optimal equilibrium. If an interaction has only one equilibrium, and this equilibrium is not P-optimal, then there must be at least one other possible outcome that is P-optimal and P-superior to the equilibrium outcome. And if all of the equilibria are P-inferior to some one other outcome, then there must be at least one P-optimal outcome that is P-superior to all of the equilibria. So we could propose as a condition that all P-optimal theories must satisfy: rational interaction must result in a P-optimal outcome P-superior to all the equilibria, if there is such an outcome.

This is not the end of the matter. We suppose that a fully developed best-reply theory must single out one particular equilibrium as the outcome of a rational interaction. A P-optimal theory could then require that the rational outcome of interaction should instead be P-optimal and P-superior to the outcome proposed by the best-reply theory. If we accept this condition, then we can conclude that in any interaction in which no equilibrium is P-optimal, every person ascribes greater utility to the P-optimal outcome than to the best-reply outcome. But this would make the P-optimal theory partially parasitic on the best-reply theory. One would need to know how the best-reply theory ranked non-optimal equilibria (in a situation lacking any P-optimal equilibrium), to know just which P-optimal outcomes could be prescribed by the P-optimal theory.

What if all or some of the equilibria are P-optimal? Then one could accept, as a condition that both best-reply and P-optimal theories must satisfy, that they select one of the outcomes that are both P-optimal and in equilibrium. The same one, if there is more than one? It would seem plausible to assume this. If we do, then we can say that if some outcome is in equilibrium and P-optimal, best-reply and P-optimal theories would converge in their selection of outcomes. Every person would expect the same utility whether he reasoned on best-reply or on P-optimal grounds. If there are P-optimal equilibria among the possible outcomes of some interaction, then both best-reply and P-optimal theories may claim to confer rationality on the same outcome, though they will offer different rationales for selecting it.

In any given interaction, the outcome selected on P-optimizing grounds will either be Pareto-superior to or identical with the outcome selected on grounds of maximizing reply. Faced with the same environment, the members of a community of P-optimizers must do at least as well, and if they ever face situations in which some person's best-reply action is not P-optimal, must do better than the members of a population of maximizers. The P-optimizers coordinate their choices one with another; the maximizers treat their fellows merely as constraints on their individual pursuit of utility. We might say that cooperators *interact* with their fellows, coordinating their choices for mutual benefit, whereas maximizers simply *react* to the choices of others in seeking their own benefit.

Maximizers are of course not unmindful of the problems they face when their best-reply choices fall short of P-optimality. They do not and cannot deny the benefits of cooperation, so they seek to attain them by imposing institutional constraints on the options available to the agents. They endeavor to introduce rewards, and more plausibly punishments, that will convert what otherwise would be Dilemma-like situations into interactions with P-optimal equilibria. The objection to these measures, at the level of theory, is evident. All enforcement devices come with a cost. This cost must be subtracted from the benefits that unforced cooperation could bring to the agents. But the cost must be paid, say the economists, because persons are not naturally cooperators. Left to themselves, they will maximize their payoffs against the choices they expect the others to make. Cooperation, on the view that individual utility-maximization is rational, must be grounded in a deeper level of non-cooperation.

I find this verges on incoherence. Persons are concerned to realize their aims, whatever their aims may be. Reason, practical reason, must surely be understood as conducive to this realization. But best-reply accounts of rationality make it frequently an enemy to be outwitted if possible. If that is so, then why suppose that best-reply reasoning embodies practical rationality?

But here we should admit a qualifying note. I have noted that for a particular individual, P-optimization is not always possible, or, if possible, acceptable. The other person or persons may choose in such a way that no Pareto-optimal outcome may be accessible. Or, as in the Prisoner's Dilemma, the other person may opt to defect, in which case the only accessible act that would yield a P-optimal outcome would be to cooperate, which is obviously unacceptable since that would yield one's worst outcome. It is only if one expects others to conform to P-optimization that one should do so oneself. And it is only if persons recognize the benefits of cooperation to them that it can be reasonable for them to conform. But the presence of benefits is not a given. In some interactions, the objectives of the persons are directly opposed, one to the other. If your gain is my loss, and your loss my gain, then best reply may seem rational.

2.5 Maximizing vs. cooperating

A complete P-optimal theory of rational choice must do at least two things. First, it must specify for each interaction structure the P-optimal outcome which rational persons should seek, and so the choice of actions, one for each person involved, that result in the favored outcome. Second, it must also provide rational guidance for persons who are interacting with others who are not guided by the theory. Presumably if the others are acting in ways close to those that result in acceptable P-optimality, a person should be guided by the theory, but he should be prepared for whatever others may do. And when other individuals are not cooperators, then best reply may be the reasonable alternative.

A best-reply theory will also have two tasks. The first is of course to specify for each interaction structure, the equilibrium outcome which persons rational by its standard should seek, and so the choice of actions, each a best reply to the others, that yield the favored outcome. But second, its more general task is to advance the thesis that rational persons should seek to maximize their utility, given their expectations of what others will choose. If each person's expectations are correct, then the actions chosen will be in equilibrium, so that the theory need only provide a way of choosing among equilibria. But if, as is likely, some persons, whatever their intentions, fail to make best-reply choices, then the theory must provide guidance to the others.

We might be led by considerations such as these to think of best-reply reasoning as basic, so that it constitutes the default conception of rational interaction. The status of P-optimal reasoning would then be deeply problematic. But I do not think that we should accept this proposal. Human beings are social animals not well equipped for solitary existence. Their reasoning about what to choose reflects this sociability. But human beings are also distinct individuals, with interests and concerns that are not identical with those of their fellows. Their reasoning about what to choose reflects this individuality. That is why neither individual utility maximization nor collective maximization (the utilitarian view) offers a correct account of rational interaction. The alternative is cooperation, which offers a social framework for individual choice.

But if this is at the root of my argument, it abstracts from the details that are afforded by the Prisoner's Dilemma. The best-reply theorist finds that in the Prisoner's Dilemma rational action, as he understands it, leads to a P-inferior outcome. I want now to show in detail why we should not accept this account of the Prisoner's Dilemma.

2.6 Prisoner's Dilemma and Builder's Choice

First, I want to contrast the Prisoner's Dilemma (Figure 2.1) with a very similar interaction (Figure 2.2). Let us use our earlier formulation of a quite

		COL	
		Builds	Doesn't
ROW	Builds	3,3	0,4
	Doesn't	4,0	1,1

Figure 2.1 Prisoner's Dilemma

		COL	
		Builds	Doesn't
ROW	Builds	3,3	4,0
	Doesn't	0,4	1,1

Figure 2.2 Builder's Choice

typical example of the Prisoner's Dilemma, one in which the gap between second-best and third-best is larger than the gap between best and second, or worst and third.

The story line in Figure 2.2 concerns builders rather than prisoners. Two persons who own adjacent parcels of land are each considering whether to build a hotel on his parcel. If either builds, the taxable value of all of the land will increase. But each hotel will provide its owner with increased revenue in excess of the increased expenses and taxation, and so will be profitable to him. If only one builds, he will enjoy a monopoly and profit even more. But since the taxable value of all of the land will increase, the other will be worse off than if neither had built. It is evident that each should build, yielding the mutual second-best outcome.

What distinguishes these two interactions? If each of the persons assigns his least favored outcome the value 0, and his most favored the value 4, then the other two possible outcomes will take values between 0 and 4; for our example we shall suppose them to take the values 1 and 3. If we plot the possible outcomes in utility space, we can select the same four points for each. Further, let us suppose that Row and Column could (should they so wish) randomize at will over their two actions. Then the possible expected outcomes will correspond to all of the points in utility space that fall on or within the lines that join the outcome points. Thus the two interactions share the same region of utility space. Nevertheless, best-reply theories select the third-best outcome in the Prisoner's Dilemma, but the second-best in Builder's Choice.

The difference between these interactions is then not in what is possible, since the possible outcomes are the same, but in the way these possibilities are linked. In the Prisoner's Dilemma, they are linked so that each can assure himself only of his third-best outcome. In Builder's Choice, with different linkage, each can assure himself of at least his second-best outcome. But the agents are concerned with what they get, not with what they can assure themselves. They can both get their second-best outcome in both situations, and neither can improve on this without the other losing out. The Prisoner's Dilemma and the Builder's Choice differ in their strategic structure. But they do not differ in their payoff structure. And it is the payoff structure rather than the strategic structure that matters to the persons involved. How a particular set of net payoffs is reached is of no importance.

What may blind us to this truth is that we often think of an objective as distinct from the means by which it is brought about. And of course these means may have value for us over and above that of the objective. So when we compute the utility of some course of action, we must take both means and objectives into account. And once that has been done, it would be double counting to reintroduce means once more into the calculation. So unless persons ascribe a value to a course of action just because it is a best reply, it is a matter of indifference to the agent whether an outcome is reached by choosing his best reply to the actions of one's fellows, or in some other way – such as P-optimizing. For I do not suppose that P-optimizing has value over and above the payoff it affords. Although, since it is a visible sign of our willingness to engage cooperative interaction, it has more to be said for it than can be said for utility maximization.

The Prisoner's Dilemma and the Builder's Choice offer similar opportunities and should yield similar outcomes. That game-theoretic analysis does not prescribe choices that lead to similar outcomes would be no criticism of the analysis, if it were understood as showing how acting in accordance with a particular set of rules can lead to quite different outcomes in the Prisoner's Dilemma and the Builder's Choice, despite their similarity. Suppose ourselves in a game where each person seeks to play his best reply to what the others play. If each succeeds, then the outcome of the game will be an equilibrium. Economists and others may think that this game models real life. But it doesn't. The objective in the game, to achieve a best reply outcome, does not model the objectives persons have in real life. Those objectives are given by their utilities. The success of their deliberations is found in the payoffs they receive. Cooperators win in Prisoner's Dilemma interactions. Maximizers lose.

2.7 Temporal extension of the Dilemma

To this point I have focused on interactions in which the persons effectively choose simultaneously. But many interactions are extended in time. Let us then consider the temporal extension of the Prisoner's Dilemma. Unlike the simultaneous Prisoner's Dilemma, in which each chooses in ignorance of the other's actual choice, one person, Row, chooses first, so that when Column chooses, he knows what Row has done. The story line runs like this. Row and Column are farmers. Column's crops are ready for harvesting the week before Row's are ready. Column tells Row that he, Column, will help Row next week, if Row will help him this week. The interaction is best represented in a tree diagram; see Figure 2.3.

Both Row and Column would prefer to harvest together than to harvest alone. But each would most like to receive aid in his harvesting without reciprocating, and each would of course like least to help the other harvest without receiving aid in his own harvesting.

Row is about to accept Column's offer when a doubt crosses his mind. "If I help Column this week, he'll have what he wants, so what can he gain from helping me next week?" But the doubt may seem erased, when Row reflects that were Column to go back on his offer to help Row, in future years Column could not expect any assistance from Row. And we may suppose that he would gain less from reneging on his offer than he could expect to lose in getting no assistance from Row in future years.

The tree diagram treats this year's harvesting as self-contained, ignoring the payoffs of future years. So the story needs to be modified. Column has decided to sell his farm after the harvest, and retire to Mexico, where he is most

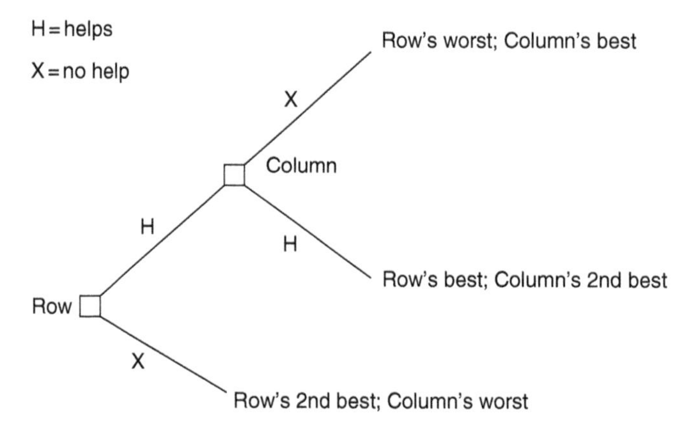

Figure 2.3 Backwards induction

unlikely to encounter Row or anyone else who might know of Column's perfidy. The tree diagram now correctly represents the interaction and its payoffs. Column reasons that whatever Row does – help with Column's harvesting or not – he, Column, has nothing to gain from helping Row. So his offer to Row is not genuine, since he knows that when the time comes, his best course of action would be not to assist Row, reneging on his offer if Row has acted on it.

Row of course notes that whatever he does – help Column or not – Column's best course of action will be not to assist him. So Row does not accept Column's offer. He chooses not to help Column, which of course ensures that Column will not help him. And so Row and Column harvest alone, although both of them would have benefited had they harvested together. If sound reasoning leads Row and Column to harvest alone, then they are victims of their rationality.

So is the reasoning that leads Row and Column sound? It runs from last choice to first choice. The last person to choose (Row in our example) surveys the possibilities that are left to him, and selects the one maximizing his own payoff. This may be far from the outcome he would most have preferred, but that outcome may have been eliminated by earlier choices. We assume full knowledge of options and payoffs, so that the penultimate chooser (Column) can determine what Row would choose for each of the options he (Column) might leave him. But then Column is in a position to evaluate these options from his own point of view, and choose to leave to Row the option that will result in Row's choosing the largest payoff for Column. Row is, of course, interested only in his own payoffs, but Column, by his choice, can direct Row's interest so that his choice yields Column a greater payoff than would any alternative open to Row.

I have examined a two-person one-move situation. (I shall discuss a multi-move situation presently.) But it exhibits what is essential in the extended Dilemma. The assumption, made among rational-choice theorists, is that at every choice point, the chooser will look to the effect of his choice on his future utility. He considers only what he can bring about; the past cannot be changed by what he does now. I was traveling by car to a conference, one of whose main themes was the logic of deterrence, a subject that overlaps rational choice. For want of something better, the car radio was tuned to one of those American stations that plays "golden oldies." And what should come across the air waves but someone singing: "Don't stop thinking about tomorrow / Yesterday's gone, yesterday's gone." This was not what I would have hoped to hear.

Yesterday's gone, but what happened yesterday may have been affected by what is expected to happen today or in the future. I want to argue that it is rational for a person to be willing to perform his part in an agreement, provided the other(s) can be expected to do, or have already done, their part. This is central to the idea of rational cooperation. A rational cooperator seeks to bring about cooperative outcomes, outcomes that are advantageous to everyone in the cooperating group. The temporal ordering of the actions he performs to achieve mutual benefit should not in itself be of concern to a rational person. In the example of our harvesters, aiding each other is P-optimal; harvesting alone is a best-reply outcome, Pareto-inferior to harvesting together. That harvesting together requires one farmer to make his contribution after he has received the benefit he sought should be no more problematic than any other feature that can be shown to enhance cooperation.

The farmers want to cooperate. They want to cooperate in harvesting this year, whether or not they will find other circumstances for cooperating. They do not need to ground their cooperation in some enforcement device that will make it costly for the person who chooses last to look only to his own future benefits in harvesting. They do not need to appeal to any supposed moral feelings that will override their payoff values. Moral feelings may be expected to favor cooperating, but this is because these feelings are derived from and supportive of the social structure, which is justified as providing the framework for cooperation.[8] The orthodox economist has to argue that it is not rational for Row to assist Column in harvesting, despite his having benefited from Column's assistance, unless there is some further, future benefit he can attain by assisting, or there is some exogenous cost imposed on his not assisting.

2.8 Coordination and commitment

I am not arguing that cooperation is always possible or, even if possible, rational. And in dealing with non-cooperators, one should fall back on maximizing one's own utility, since no better outcome can be achieved. What I am arguing is that if cooperation is – and is recognized by (most) other persons to be – possible and desirable, then it is rational for each to be a cooperator. Cooperation is desirable when the objectives of persons can be

[8] This point should be addressed in a paper of its own. Here I assume it dogmatically.

more effectively realized by them acting to bring about some agreed outcome, than by each choosing for himself on the basis of his expectation of others' choices. It is possible when its desirability is recognized, and it is then rational for persons to think of themselves as cooperators and to deliberate together about what to do.

Deliberating cooperatively involves two quite different aspects. Individual cooperators, deliberating in a particular situation, look to determine if there is a salient P-optimal outcome. If so, and if they believe themselves to be among cooperators, they look for the route to this outcome, and do what will best move their fellows and themselves along this route. In the Prisoner's Dilemma, this is simply to cooperate. In the farmers' choice, they assist one another in harvesting, regardless of who harvests first. But these are of course very simple interactions in which there is one obviously salient P-optimal outcome that they can easily choose to bring about.

The other feature of cooperative deliberation that I want briefly to mention is addressed to a very different problem. Frequently there is no salient P-optimal outcome, but a range of such outcomes. Coordination typically does not occur spontaneously, but requires the conscious agreement of the persons involved in the interaction for each to act so that the outcome will be P-optimal, and also will afford each person an equitable share of the benefits of cooperation. If this agreement can be obtained, then cooperation can be made determinate for the interacting persons.

In all but the simplest cases, cooperation either requires or is enhanced by social practices and institutions. This is not a theme to be pursued in this essay, but it is important to note that the role ascribed to the social system is primarily coordinative. It tells persons what they should do to reach an equitable P-optimal outcome. This is not a descriptive claim about actual societies, but a normative claim about acceptable societies. Or at least it endeavors to be such. For even if our society were to aim at fair and mutually beneficial interaction, our knowledge may not be sufficient to the task of discovering paths from here to P-optimality.

Coordination contrasts with constraint. In the Prisoner's Dilemma, persons who consider it rational to maximize will need to be constrained in some way that diminishes the appeal or the availability of the default option. P-optimizers do not need any external constraint to bring about their favored outcome. That best-reply agents do not find it rational to perform the actions needed to achieve their favored outcome seems to me sufficient reason to question their view of practical rationality. I shall turn shortly to build on this criticism, and indeed offer a *reductio ad absurdum* argument by examining the implications of their prescription in one more scenario – the iterated Prisoner's Dilemma.

I do not suppose that in a situation in which no best reply is P-optimal, one will not be tempted to depart from the P-optimal path especially if one is being told (by the economists) that it is rational to do so. I have suggested that moral feelings will support cooperation, but Pareto-optimality is not prominent in our actual moral vocabulary, so that moral support for P-optimal actions may be relatively weak. But the way to provide such support most effectively is, I think, not through external constraint, but through internal constraint embedded in the idea of commitment.

Commitment is necessarily misunderstood by most economists and game theorists – necessarily, because commitment involves a rearrangement of deliberative space without altering either the utilities or the circumstances of the persons involved. Orthodoxy requires that if a person expresses an intention to choose a particular action in the future, but then, when the time for choice comes, that action is not her best reply or does not maximize her expected utility, she must, as a rational agent abandon her intention and choose what now seems best. Commitment denies this. If a person commits herself to a future choice, she expresses an intention that she will not abandon simply because it proves non-maximizing. As long as honoring the commitment makes her payoff greater than had she not formed and expressed it, she has sufficient reason to honor it. And even if honoring a particular commitment may leave her worse off than had she not made it, if she commits herself in accordance with a policy of commitment that is beneficial for her, then she has sufficient reason to honor the commitment.[9]

We want to be able to assure others that we will make certain choices even if it is costly to do so. Row wants to assure Column of his assistance, even though he will have no future-oriented reason to fulfill it. So he promises Row his assistance. Promising is one of the standard ways we have of creating a commitment. Row being committed, Column is willing to join in harvesting Row's crops this week, because he is assured of Row's participation next week.

The difference between the view of commitment I am suggesting and the view held by such game theorists as Ken Binmore should be evident. "If it is claimed that commitments can be made, game theorists want to know what the enforcement mechanism is. If a convincing enforcement mechanism is described, then the opportunity for making a commitment can be included

[9] See Gauthier (1994: 717–19) for some discussion relevant to this point.

as a move in the game."[10] This is exactly what Binmore must say, but it does not capture our ordinary understanding of commitment. A commitment does not affect the actions available to the agents, or the utilities afforded by these actions; it does affect what he takes to be his reasons for acting.

To continue, I should have to propose a theory of rational commitment that dovetails with a P-optimizing account of rational deliberation. But that would be a task for another essay. So I shall put commitment to one side, and turn to the problem, for maximizing accounts of practical rationality, posed by the iterated Prisoner's Dilemma.

2.9 Iterated Dilemmas

Suppose two persons face a series of interactions, each of which, considered in itself, would be a Prisoner's Dilemma. In each interaction, the persons are aware of how the two of them acted in all previous interactions, and what the payoffs were. Thus each can decide, in the current situation, not only to cooperate or to defect, but, for example, to cooperate if and only if the other person cooperated in the immediately preceding interaction. The interactions are not, therefore, strictly speaking Prisoner's Dilemmas, because each may have consequences for later interactions, and these further consequences will of course affect the utilities of the agents. The true Prisoner's Dilemma is self-contained, but I shall for once follow the example of orthodox game theorists and others in continuing to speak of the Iterated Prisoner's Dilemma.

Now I have nothing original to say about determining what to do in an Iterated Prisoner's Dilemma. I want only to point to the inability of game theorists to provide an account that always satisfies the best-reply condition, and has any plausibility whatsoever as determining *rational* behavior. We might think a plausible candidate for meeting these conditions would be the one suggested in the previous paragraph – cooperate in the initial Prisoner's Dilemma, and thereafter choose in Prisoner's Dilemma n what the other person chose in Prisoner's Dilemma $n-1$. But alas this proposal does not satisfy best reply. For it is clear that when the last member of the sequence of interactions is reached, a person's best reply, whatever the other chooses, is to defect. And if this is so, then one knows that the other will defect, since our theory of rational choice must give each person in a symmetrical interaction the same prescription. But then in the penultimate interaction, one's

[10] Binmore (1991: 140).

choice can have no effect on the other's final choice, and so one does best to defect. And, alas, so on, until one reaches the initial interaction, and concludes that one should defect from the outset. There are of course other proposals for how one should act in the Iterated Prisoner's Dilemma, but, to quote Luce and Raiffa, "any equilibrium pair of strategies will result in the repeated selection of [defect] on each of the [Prisoner's Dilemmas]."[11]

I know – game theorists do not find this conclusion palatable, and so seek out plausible variations on iteration that do not give rise to the problem of terminal defection, or propose plausible sounding courses of action that flagrantly ignore equilibrium considerations. If the agents do not know the number of iterations, so that any might be the final one, cooperation might be sustained by best-reply reasoning. But the bottom line is that the Iterated Prisoner's Dilemma, in its purest and simplest form, shows that best-reply theories must endorse as rational interactions that in our intuitive understanding would be the height of folly. And this I take to be justification for seeking an alternative account of rationality.

A P-optimal theory of practical rationality makes short work of situations like the Iterated Prisoner's Dilemma. But it does so by invoking the Pareto-condition on practically rational outcomes. How do we know that this is the correct condition? We can test it and its rivals against our pre-theoretical judgments of what it makes sense to do. We can compare its deliverances with those of its rivals. We can compare its structure with the structures of other normative theories. But the question of correctness must be left to another occasion.

2.10 The Cooperators' Opportunity

So let us return one more time to the pure Prisoner's Dilemma. Far from fitting the label "dilemma," the Prisoner's Dilemma exhibits the clash between two rival accounts of practical rationality. We can always have an outcome in equilibrium or we can always have one that is P-optimal, but we cannot always have both. Much of economic theory focuses on interactions in which the two coincide. When they do, it is as if each person intends only his own benefit, but a benign Invisible Hand leads them to benefit all. Unfortunately, if in the Prisoner's Dilemma each person intends only his own benefit, a malign Invisible Hand leads them to benefit none.

[11] Luce and Raiffa (1957: 99).

A P-optimal approach, enabling persons to cooperate, whether by coordinating their actions, or even relating their actions to a single directive addressed to all, employs no Invisible Hand. It treats persons as aiming at mutual benefit, in any context in which their objectives can be best achieved if each does not intend only his own benefit. And this, of course, is the situation of the Prisoner's Dilemma. When I first encountered the Prisoner's Dilemma, I was unconvinced by those who thought mutual defection was the rational way to act. But I saw the Dilemma as a problem – how to introduce non-maximizing behavior into a maximizing framework? Put this way gives utility-maximization more than its due. The Prisoner's Dilemma reveals two ways in which interactions might be assessed as rational, the familiar maximizing account, and the Pareto-optimizing account. It shows that in some, very basic interactions the two yield incompatible directives, and it shows the superiority of P-optimizing to maximizing by a standard that both accounts accept – the payoffs that result from following the directives. True it is that *we* must P-optimize; there is no going it alone. And that may be a practical problem. But only a practical problem. Rightly understood, the Prisoner's Dilemma shows the way to achieve beneficial interaction – and the way not to achieve it. We should relabel it – the Cooperators' Opportunity. And that is how I learned to stop worrying, and love the Prisoner's Dilemma.

3 Taking the Prisoner's Dilemma seriously: what can we learn from a trivial game?

Daniel M. Hausman

The Prisoner's Dilemma in my title refers to a one-shot two-person game with the extensive form shown in Figure 3.1:[1]

In Figure 3.1, player 1 chooses C or D. Player 2 does not know which was chosen. The two different nodes where player 2 chooses are contained within a single information set, represented by the dotted oval. The pairs of numbers at the terminal nodes represent the preferences of respectively player 1 and player 2, with larger numbers indicating outcomes that are higher in their separate preference rankings. Whenever I refer to the Prisoner's Dilemma, it is this game that I am talking about. I have followed convention in labeling the two strategies of the two players "C" and "D," which suggests that C is a strategy of cooperation and D a strategy of defection. Because the strategy choice of player 1 is contained within a single information set, the game is equivalent to one involving simultaneous play. It makes no difference whether player 1 is depicted as moving first or player 2 is depicted as moving first. The normal or strategic form shown in Figure 3.2 represents the game more compactly.

The payoffs are ordinal utilities – that is, indicators of preference ordering. The first number in each pair indicates the preference ranking of player 1, while the second number indicates the preference ranking of player 2. These numbers only indicate the ranking. They have no other significance. So, for example, if one were to substitute 12 for one of the 4s in either representation of the game, 12,468 for the other, and −36 for both of the 1s, it would make no difference to the game.

As I have argued elsewhere, the preferences that are indicated by the utilities in games are total subjective comparative evaluations. What this means is the following:

[1] I am grateful to Reuben Stern for comments and suggestions. Parts of this essay derive from my *Preference, Value, Choice and Welfare* (2012), especially chapter 6.

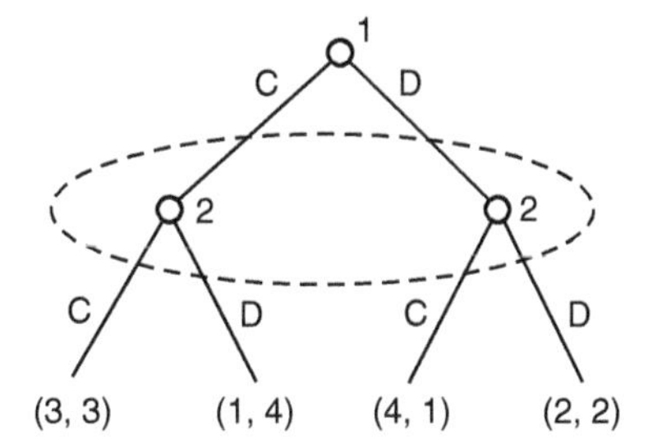

Figure 3.1 Prisoner's Dilemma: Extensive form

		2: C	2: D
1	C	3, 3	1, 4
	D	4, 1	2, 2

Figure 3.2 Prisoner's Dilemma: Normal form

1. Preferences are *subjective states* that combine with beliefs to explain choices. They cannot be defined by behavior, contrary to what revealed preference theorists maintain. Just consider the Prisoner's Dilemma game, which represents player 1 as preferring the outcomes where player 2 chooses strategy C to the outcomes where player 2 chooses strategy D. These preferences of player 1 cannot be defined by any choices of player 1. If, instead of attempting to define preferences by choices, one attempts to avoid references to subjective preferences altogether, then one cannot represent the game. Even in the simplest case in which an individual faces a choice between just two alternatives, x and y, the choice of x might result from the agent's falsely believing that y was not available rather than from any disposition to choose x over y. One cannot draw conclusions about either preferences or future choices from observed choices without premises about beliefs.
2. Preferences are *comparative evaluations*. To prefer x to y is to judge how good x is in some regard as compared to y. To speak of a preference for x is elliptical. To prefer x is to prefer it to something else.
3. Preferences are *total* comparative evaluations – that is, comparative evaluations with respect to every factor that the individual considers relevant. Unlike everyday language, where people regard factors such as moral duties as competing with preferences in determining choices, game theorists take preferences to encompass every factor influencing choices, other than beliefs and constraints.

4. As *total comparative evaluations*, preferences are cognitively demanding. Like judgments, they may be well or poorly supported and correct or incorrect. There were good reasons to prefer that Romney be elected. The possibility that an asteroid may strike the earth was not one of them.

As others have argued,[2] strategy choice in the Prisoner's Dilemma is trivial: D is a strongly dominant strategy for each player. Provided that (1) the players know that they are playing a game with the structure of Prisoner's Dilemma, (2) the players understand that regardless of what the other player does, they prefer the outcome of playing D, and (3) the choices of the players are determined by their preferences, both players will play strategy D.

Two facts make this trivial conclusion interesting. First, both players prefer the result of a pair of strategies they do not choose – that is the result of (C, C) – to the result of the strategies they do choose. So this simple game conclusively refutes a naive exaggeration of Adam Smith's claims about the benefits of the invisible hand. The Prisoner's Dilemma shows that there are some institutional frameworks in which the self-interested choices of rational individuals are not socially beneficial. It warns us that individually rational choices can have socially harmful consequences. Whether rational choices will have beneficial or harmful consequences in a particular market setting requires investigation. If some market institutions guarantee the social optimality of self-interested individual choices, it is because of properties of those markets, not because there is something automatically socially beneficial about the rational pursuit of one's interests.

Second, when experimenters confront experimental subjects with strategic problems that appear to be Prisoner's Dilemmas, many of them play strategy C. For example, an affluent experimenter may present experimental subjects (whose identities are not known to each other) with the strategic situation shown in see Figure 3.3.

Call situations such as these "money-Prisoner's Dilemmas." The subjects are told that the circumstances and payoffs are common knowledge, and each is asked to choose a strategy without knowing what the other will choose. Anonymity is crucial to block the influence of future uncontrolled interactions between the experimental subjects. Though a majority of subjects in money-Prisoner's Dilemmas typically choose D, many choose C.

[2] I have in mind especially Ken Binmore (1994), whose compelling arguments are independent of his untenable commitment to revealed preference theory.

		2	
		C	D
1	C	\$30, \$30	\$10, \$40
	D	\$40, \$10	\$20, \$20

Figure 3.3 A "money-Prisoner's Dilemma"

According to game theory, C is obviously the wrong choice for players who care exclusively about their own monetary payoffs. Why then do so many people make this choice? Perhaps people are just dumb. Or maybe there is something fundamentally wrong with game theory. A third possibility is that people do not care only about their own monetary payoffs. People are in fact sometimes less than entirely clear-headed, and some experimental work apparently supports the confusion hypothesis. In particular, if experimental subjects play money-Prisoner's Dilemmas (or very similar collective goods games) repeatedly against different people (so that they are still playing one-shot games), the frequency of defection increases until only a small percentage play C (Andreoni 1988, Fehr and Gächter 2000). But if, after playing several rounds of money-Prisoner's Dilemma games, experimental subjects are told that they are starting over and playing a new set of rounds (again against different players each round), then many begin, as before, by playing cooperatively (Andreoni 1988). These data do not rule out an explanation in terms of confusion (and learning), but they render it implausible.

A second possibility is to argue for a revision of game theory or an alternative way of applying game theory to the Prisoner's Dilemma. As Binmore argues (1994, chapter 3), most of these arguments are confused. But there are plausible proposals for reinterpreting game theory. Amartya Sen makes the case that rational decision-making is not just a matter of picking one of the feasible alternatives at the top of one's preference ranking. In addition, people's choices are influenced by "commitment," which is not self-interested (1977) and, Sen argues, can be contrary to preference. Although I shall argue that Sen's concerns can be met without surrendering the standard view of preferences as total comparative evaluations, let us for the moment think of preferences, as Sen does, i.e., as not exhausting all of the factors that may be relevant to choice. I shall refer to rankings based on only some of the relevant considerations, which Sen calls "preferences," as "preferences*." It is possible for rational agents to choose one strategy even though they prefer* the outcomes of another.

For example, suppose the utilities in the Prisoner's Dilemma game indicate a *partial* ranking in terms of expected self-interested benefit rather than a total ranking in terms of all factors that the players take to be relevant to their choice. If other considerations apart from expected advantage influence the choices of the players, then this interpretation of the payoffs makes room for counter-preferential rational choice. If other things matter, then the fact that D is a dominant strategy in terms of expected advantage does not imply that rational individuals will play D. Sen's view opens a space in which to discuss the thought processes that may lead individuals to make choices that are not governed by their expected advantage or partial preferences. There are many possible routes that might lead individuals rationally to choose C, despite having a preference* for D. For example, individuals faced with the money payoffs shown in Figure 3.3 might think to themselves:

> Whatever the other player does, I get more money by playing D. So that is where my expected benefit and preferences* lie. But my choices are not governed only by my preferences*. I also care about being a decent person, and if I think the other player will play C, then reciprocity rules out strategy D. I think it is common knowledge that most people are pretty decent and, like me, most will pass up the chance of getting what they most prefer* when they believe that the result of doing so is a high equal result, like our both getting $30. So I think it is common knowledge that most people, like me, will choose C. Although I'm passing up the outcome I most prefer*, it's not that costly to do the right thing. So I choose C.

It is not far-fetched to suppose that some people think this way, and indeed if one attributes reasoning like this to experimental subjects, then one has a ready explanation for the decay in cooperation that occurs when people play money-Prisoner's Dilemmas repeatedly: they discover that they were overly optimistic about the choices of others. When they no longer believe that others are likely to play C, they face the choice between a suboptimal but equal result (where both players get $20) and an unequal result in which they are the loser, getting only $10, while the other player gets $40. Decency or reciprocity does not require that they do worse. When the game is restarted, their confidence that others will play cooperatively increases and hence the level of cooperation increases. It thus appears that there may be a significant gain in interpreting preferences more narrowly than as total comparative evaluations.

Sen's argument that cooperation can be rational in one-shot Prisoner's Dilemmas does not rest on a specific interpretation of preferences in terms of self-interested expected advantage. Indeed, he notes that altruists can find themselves in a Prisoner's Dilemma, too. What is crucial to Sen's approach is his denial that preferences are total comparative evaluations: as he understands preferences, they do not incorporate all of the considerations bearing on choice. He writes:

> The language of game theory... makes it... very tempting to think that whatever a person may appear to be maximizing, on a simple interpretation, must be that person's goal... There is a genuine ambiguity here when the instrumental value of certain *social* rules are accepted for the *general* pursuit of *individual* goals. If reciprocity is not taken to be intrinsically important, but instrumentally so, and that recognition is given expression in actual reciprocal behaviour, for achieving each person's own goals better, it is hard to argue that the person's "real goal" is to follow reciprocity rather than their respective actual goals. (1987, pp. 85–6).

I interpret these complicated remarks as follows: Suppose the utilities that define a Prisoner's Dilemma game indicate only *intrinsic* evaluative considerations, such as how much money one gets or one's moral commitments. They are not influenced by instrumental concerns about trustworthiness or reciprocity that can contribute to achieving what people intrinsically value. If the player's actions are influenced by considerations such as collectively instrumental reciprocity or trustworthiness, then it can be rational for them to play C, despite the fact that C is worse in terms of intrinsic evaluative considerations. As Sen maintains, when the reciprocity the players show is instrumental to pursuit of what they value intrinsically, "it is hard to argue that the person's 'real goal' [or preference] is to follow reciprocity rather than their respective actual goals." I take this to mean that the utilities in the game indicate only "actual goals."

Sen is right about the importance of modeling the intricate thought processes individuals go through when faced with strategic problems like the money-Prisoner's Dilemma, and his suggestion that those who decide to cooperate may still prefer* the results where they defect to the results where they cooperate is plausible. ("Sure I'd rather have had $40, but ...") Sen's approach permits one to model a wider range of thought-processes that lead to actions in strategic situations such as money-Prisoner's Dilemmas than do standard game-theoretical treatments.

When one asks game theorists why so many individuals facing a money-Prisoner's Dilemma cooperate, they should have something better to say than, "That's not our department. Go and talk to the social psychologists." Sen's proposal that economists adopt a multiplicity of notions of preference corresponding to different specific evaluative concerns and interpret the payoffs of games so that strategy choices are no longer deducible from either extensive forms such as Figure 3.1 or normal forms such as Figure 3.2 is one way to address the complexities of the players' interpretations of the strategic situation.

It seems to me, however, that the costs of divorcing choice from preference and belief are too high. If preference and belief do not determine choice, what will? In my view, preserving the possibility of predicting what players will choose from the specification of the game they are playing provides a decisive reason to take preferences to be total rankings.

Sen's concerns remain and should be addressed. A better way to do so is to conclude that game theory addresses only one part of the problem of predicting and explaining strategic behavior. In addition to game theory, economists need models of how players facing a strategic interaction *construct* their preferences and expectations and hence the game they are playing. If economists stick to the standard interpretation of preferences as total comparative evaluations – as I maintain they should – then the only plausible reading of the experimental results is that experimental subjects care about other things in addition to their own monetary payoffs: money-Prisoner's Dilemmas are often not Prisoner's Dilemmas. The task then is to determine what games experimental subjects take themselves to be playing.

It is not easy to say what game they are playing, and there are many possibilities. Consider the following restatement of the line of thought I presented above in motivating Sen's position:

> Whatever the other player does, I get more money by playing D. So that is where my expected benefit lies, and if I cared only about my own monetary payoff, I would prefer to play D. But I also care about being a decent person, and if I think the other player will play C, then reciprocity rules out strategy D. I think it is common knowledge that most people are pretty decent and, like me, most prefer outcomes like the one where we both get $30. So I think that most people will choose C. Since the outcome where we both get $30 is what I most prefer, I prefer C to D. So I choose C.

This restatement permits the same consideration of trade-offs between gains for oneself and the demands of reciprocity, conditional on expectations concerning how other players will evaluate the outcomes. I do not mean to suggest that this restatement captures the thinking of many, let alone most, of those who play C. Determining how players deliberate requires extensive inquiry, not an off-the-cuff speculation. Such inquiries are essential: there is little point to game theory if one has no idea what game people are playing or if one misidentifies the game by falsely supposing that individuals care only about their own monetary payoffs.

There have been some efforts to develop explicit general models of ways in which players interpret strategic situations and construct preferences that depend on features of the game in addition to the player's own monetary payoff. For example, Matthew Rabin (1993) provides an account whereby preferences among outcomes depend both on one's own payoff and on a concern to repay kindness with kindness and unkindness with unkindness. Ernst Fehr and Klaus Schmidt (1999) represent the influence of factors in addition to one's own payoff in terms of players' aversion to inequality, both in their favor and especially to inequalities that favor the other player. Bolton and Ockenfels (2000) model preferences as depending both on absolute and relative payoff. Cristina Bicchieri (2006) presents a model that incorporates social norms into the preferences of agents.

To provide some sense of this work, I shall sketch Bicchieri's model. In her account, a strategy profile renders a norm applicable to an individual *j* if there is a norm specifying what *j* should do in the game given the strategies that the others actually follow. A strategy profile violates a norm if for at least one player *j* there is a norm applicable to *j* that *j* doesn't obey. Bicchieri's norm-inclusive utility function is as follows:

$$\mathrm{U_j}(s) = \pi_\mathrm{j}(s) - k_\mathrm{j}\pi_{ml}(s).$$

I have changed some of her notation. $\pi_j(s)$ is what *j*'s payoff would be in the game, if *j* were not influenced by any social norms. k_j is a number which reflects how *sensitive j* is to the relevant social norm. $\pi_{ml}(s)$ is the maximum loss to any individual resulting from any norm violations by individual strategies in the strategy profile *s*, but never less than zero. How badly someone violates a norm is measured by the maximum impact the norm violation has on any player's preferences. If there is no loss to the norm violation or no norm violation, then $\mathrm{U_j}(s) = \pi_\mathrm{j}(s)$. k_j is specific to the particular agent *I* and situation. Agents may be intensely concerned to adhere to some norms in some situations and unconcerned to adhere to norms in other situations. Since Bicchieri multiplies $\pi_{ml}(s)$ by a real number and subtracts the product from $\pi_\mathrm{j}(s)$, these cannot, of course, be merely ordinal indicators of preferences.

To illustrate how this applies to Prisoner's Dilemmas, suppose there is a social norm calling for cooperation in money-Prisoner's Dilemmas. Without the norms, the players would confront a genuine Prisoner's Dilemma like the one shown in Figure 3.2. If both play C, then the second term in Bicchieri's

		2	
		C	D
1	C	3, 3	$1-2k_j$, $4-2k_k$
	D	$4-2k_j$, $1-2k_k$	2, 2

Figure 3.4 Social norms on a money-Prisoner's Dilemma

		2	
		C	D
1	C	3, 3	−1, 2
	D	2, −1	2, 2

Figure 3.5 A transformation of a money-Prisoner's Dilemma

utility function for both is zero, and so the (utility) payoff for the strategy pair (C, C) is just as it would be if there were no norms. If Jill (Player 1) plays C and Jack (Player 2) violates the norm and plays D, then Jill's utility is $1 - 2k_j$ and Jack's is $4 - 2k_k$. Similarly, if Jill plays D and Jack plays C, Jill's utility is $4 - 2k_j$ and Jack's is $1 - 2k_k$. Since no norms are applicable to Jack or Jill when both play D, the payoffs to the strategy pair (D, D) are as they would be if there were no norms. So the actual utility payoffs in the game Jill and Jack are playing are as shown in Figure 3.4.

If both k_j and k_k are very small, the players facing the money-Prisoner's Dilemma can be regarded as playing a Prisoner's Dilemma game, despite the inconsequential possibility of taking the payoffs as representing more than merely the ordering of preferences. If both k_j and k_k are greater than or equal to one, then this is a coordination game in which cooperation is rational. If, for example, k_j and k_k were both equal to one, then the players would be playing the coordination game shown in Figure 3.5.

Since the payoffs in Figure 3.5 are calculated from $\pi_1(C)$, $\pi_1(D)$, $\pi_2(C)$, and $\pi_2(D)$, these cannot be merely representations of what the preference ordering would be in the absence of social norms. The values of $\pi_1(C)$, $\pi_1(D)$, $\pi_2(C)$, and $\pi_2(D)$ must be unique up to a positive affine transformation. But one can regard this as an idealization that permits us to represent the processes where the preference ordering of the players is determined. The payoffs in the normal form in Figure 3.5 can still be regarded as representing nothing more than the ordering of preferences.

I am not defending the details of Bicchieri's proposal, which I have elsewhere criticized (2008). The discussion here aims only to illustrate how economists might accommodate Sen's concerns without surrendering their conception of preferences as total comparative evaluations. As already mentioned, there are other prominent alternatives in the literature. It is important not to conflate the distinct factors that may cause people's overall ranking of outcomes to diverge from the ranking in terms of their own monetary payoffs. One should distinguish among at least the following five factors: altruism, reciprocation, inequality aversion, social norms, and trustworthiness. In addition, the salience and force of these considerations depend upon the expectations players have concerning what others will do and on interpretations of the actions of others (in games unlike the Prisoner's Dilemma where players can observe the moves of others before they have to act).

To argue that economists should seek an explicit theory of how games are constituted, which would include an account of how individuals in a strategic interaction construct their beliefs and preferences, does not require arguing that economists break the connection between dominant strategies and rational choices. The way to win Sen's battle for greater sensitivity to the complexities of deliberation is for economists to model the process through which strategic circumstances become games, rather than to reconceptualize games themselves, which is what rejecting the concept of preferences as total subjective comparative evaluations requires.

Although there are, no doubt, some people who cannot understand dominance reasoning or who just get fuddled when asked to make a choice, the data suggest that what explains why people cooperate so frequently in money-Prisoner's Dilemmas is that appearances are deceptive: they are not playing Prisoner's Dilemma games. Since they do not care only about their monetary payoffs, defection is not a dominant strategy. But to say this is not to say what games people are playing or to predict how they will choose. In addressing these further tasks, it would be foolish to limit one's inquiry to money-Prisoner's Dilemmas or to comparisons of actual games to genuine Prisoner's Dilemmas. But the Prisoner's Dilemma provides one simple way of seeing the limitations of game theory as a framework for the explanation and prediction of people's choices and the need for explicit modeling of the ways in which people who confront strategic situations construct games.

4 Prisoner's Dilemma doesn't explain much

Robert Northcott and Anna Alexandrova

4.1 Introduction

The influence of the Prisoner's Dilemma on economics, law, political science, sociology, and even anthropology and biology is hard to overstate. According to JSTOR, almost 16,000 articles about it have appeared since 1960, with no sign of slowing down: 4,400 were just in the last 10 years. It has a high profile in non-academic media too. It appears as an explanation of phenomena as disparate as business strategy, political bargaining, gender relations and animal behavior. Historians of social science have referred to the Prisoner's Dilemma as a "mainstay" (Morgan 2012: 348) and an essential "set piece" (Rodgers 2011: 64). And according to Robert Axelrod, "the two-person iterated Prisoner's Dilemma is the *E. coli* of the social sciences" (quoted in McAdams 2009: 214).

As philosophers, our aim is to assess whether this development has been worthwhile and furthered the goals of social science. We ask this question even knowing that it cannot be answered fully in a single article. The research programs that the Prisoner's Dilemma has inspired are many and diverse, and the Prisoner's Dilemma is only one of many models that have been used in them. In addition, social science, like science in general, has many different goals and a judgment of worthwhileness requires a devilishly complex integration of conflicting considerations and values. Finally, sixty years may or may not be a sufficient span to judge. Nevertheless, we will brave giving a prima facie case that on at least one central criterion, namely providing causal explanations of field phenomena involving human co-operation, the Prisoner's Dilemma has failed to live up to its promise.

Before we start, two clarifications are in order. First, we do not wish to criticize the use of the Prisoner's Dilemma on moral or political grounds. It might be that teaching and using it makes people behave more selfishly and normalizes a narrow conception of rationality (Dupré 2001, Marwell and Ames 1981). But our concern is purely methodological: has the Prisoner's Dilemma delivered empirical success?

Second, we focus on the Prisoner's Dilemma because it is the subject of this volume, not because it is unique in the way it has been misused. Much of what we say applies to other analyses of collective action problems, and much of economic theory more generally. But here our focus will be on the Prisoner's Dilemma only. It is quite plausible that often the Prisoner's Dilemma gets misused just because it is uniquely famous, so scholars invoke it when instead they should be invoking a different game, say the Stag Hunt, or another co-ordination game (McAdams 2009). That is a mistake, but not the one we care to correct here, if only because correcting it would call for greater use of the very models that we argue do not provide a good return on investment anyway.

In Section 4.2 we present an account of how the Prisoner's Dilemma could provide causal explanations. The heart of the paper is in Section 4.3, where we make the case that in fact it has failed in this task. To this end, we examine in detail a famous purported example of Prisoner's Dilemma empirical success, namely Axelrod's analysis of WWI trench warfare, and argue that this success is greatly overstated. Further, we explain why that negative verdict is likely true generally, and not just in our case study. In Section 4.4, finally, we address some possible defenses of the Prisoner's Dilemma.

4.2 The possibility of explanation

4.2.1 What sort of explanation?

Is the Prisoner's Dilemma explanatory? There exists a canonical account of explanation known as *situational analysis* (Koertge 1975), which was originally articulated for social science by Popper, Dray, and Hempel. As Mary Morgan, among others, has pointed out, the Prisoner's Dilemma is particularly well suited to it. According to situational analysis, social scientists do not seek laws as such but rather work to define "a kind or type of event" (Popper, quoted in Morgan 2012: 358). Such a type consists in certain features of a situation, which include the circumstances (institutional, historical, environmental) of agents, plus their beliefs and desires. As a second step, one adds an analysis of what it is rational to do in these particular circumstances. The third step is the assumption that the agents are indeed rational, and then the explanation follows: a given phenomenon arises because rational agents behave thus and thus in such and such situations. Since model building in game theory follows something like this logic, the claim is that situational analysis is how these models provide explanations. Theory building on this view amounts to generating a portfolio of models which represent typical situations that arise in different domains of the social world. The Prisoner's Dilemma is one such model.

This leaves hanging an obvious question: exactly what sort of explanation does situational analysis provide? Accounts of scientific explanation abound. We will review here only the candidates most likely to apply to the Prisoner's Dilemma, without claiming any general superiority for one model of explanation over another.

If any theory of explanation can claim to be dominant in social science it is *causal* explanation. One well-known reason is its intimate connection to interventions, because interventions in turn are the lifeblood of policymaking. One prominent theory states that to give a causal explanation is to make a counterfactual claim that if a hypothetical intervention changed the cause then the effect would also be changed (Woodward 2003). We believe that something like this is the best hope for defenders of the explanatory potential of the Prisoner's Dilemma. But before returning to it, we will briefly mention two other leading possibilities.

The Prisoner's Dilemma in particular and game theory more generally is often thought to *unify* social phenomena: not just many different economic phenomena can be modeled but also political, legal, social, and personal ones too.[1] It is this unifying ambition that has earned economics more generally the accusation of imperialism. If the Prisoner's Dilemma really did unify phenomena in an explanatory way, we would welcome that and count it to the Prisoner's Dilemma's credit. But it does not. A closer look at unificationist theories of explanation, such as Kitcher's (1981), shows why. According to Kitcher, in order to explain, a theory must satisfy *two* unifying criteria: the first, roughly speaking, is range, i.e. the theory must indeed address explananda from many domains. But there is also a second criterion, which Kitcher calls *stringency.* Roughly speaking, this demands that such a unification not be vacuous – a theory must rule some things out, otherwise its compatibility with many domains is won too cheaply. Yet utility maximization, for instance, is underconstrained: utility is defined so thinly that almost anything could be an example of its maximization. This and other similar points tell against the claim that the Prisoner's Dilemma explains by unification.[2] Most likely, the needed

[1] And even sub-personal ones, as in Don Ross's game-theoretical approaches to the brain (Ross 2009).

[2] See Reiss (2012: 56–59) for more detail on why economic models do not satisfy unification theories of explanation.

constraints would have to come from causal knowledge about the contextual variation of judgment and choice, so causal explanation will return to the scene.

We believe that there is similarly no refuge in the notion of *mathematical* explanation. Perhaps, for instance, it might be thought that the Prisoner's Dilemma demonstrates the mathematical reason why two agents, given certain preferences and information and a certain environmental situation, will act in a particular way – much as statistical mechanics demonstrates the mathematical reason why with overwhelming probability heat will flow from hot air to cold. But, first, the notion of mathematical explanation of physical facts is contentious and the subject of much current debate.[3] And, second, in any case it is agreed by all that to be considered seriously mathematical explanations require empirical confirmation of precisely the kind that, we will argue, is typically absent in Prisoner's Dilemma cases.

Return now to situational analysis. This, we submit, can be thought of as an instance of causal explanation. When a social phenomenon is explained by the fact that it is an instance of a Prisoner's Dilemma, there is a claim to the effect that the structure of the situation in conjunction with the actor's rationality caused the outcome. This structure, the agents' beliefs and desires, and their rationality, are individually necessary and jointly sufficient causes of the outcome. Manipulating one of these conditions, say by changing the incentives or the information, would turn the situation into something other than a Prisoner's Dilemma and a different effect would obtain. To say that a situation is a Prisoner's Dilemma is thus just to specify a particular causal set-up.[4]

What causal structure does a Prisoner's Dilemma specify? According to Mary Morgan, there are three identity conditions for a Prisoner's Dilemma: (1) the 2-by-2 matrix which gives each player two options, (2) the inequalities that define the payoff structure, and (3) the narrative. The first two are well-known and uncontroversial, but the third ingredient is worth pausing on – what is a narrative and why would one think it essential to explanation?

As a story with a beginning, middle, and end, a narrative is the standard way of presenting the Prisoner's Dilemma. Originally at RAND, the story was

[3] See, for instance, recent work by Alan Baker, Bob Batterman, Chris Pincock, Mark Colyvan, and Otavio Bueno. For an overview, see Mancuso (2011).

[4] Admittedly, this claim runs afoul of the longstanding alternative according to which reason-based explanations cannot be causal because reasons have a normative connection to actions (e.g. Risjord 2005). If they cannot, then situational analysis is not a species of causal explanation after all. We do not wish to wade into this debate here, beyond saying that reason-based explanations routinely get re-cast as causal explanations by scientists and philosophers alike, and arguably for good reason.

Tosca's and Scarpia's attempt and failure to double-cross each other at the end of the opera *Tosca*. Later on, two prisoners' failure to cooperate against a prosecutor became the dominant story instead. Morgan insists that this storytelling aspect of situational analysis is essential but one that tends to get sidelined.[5] Yet in her view it makes the Prisoner's Dilemma what it is. First, a narrative matches the model and an actual situation – an explanandum – by providing a description of the model situation that the actual situation is supposed to match. It is thus a condition of model application. Second, a narrative provides a general category that allows for the classification of a situation as being of a particular *type*. Since situational analysis explains precisely by specifying a situation's type, the narrative is thus also essential to explanation (Morgan 2012: 362–363).

We think Morgan is right that the narrative does the explaining that the matrix and inequalities alone cannot. If she isn't right, then the whole story about situational analysis has to be abandoned too. A narrative tells us how informational and institutional constraints made agents behave as they did by providing reasons for them to behave as they did. If these constraints had been different, the agents would have behaved differently. So the narrative is essential to the explaining.[6,7]

An independent motivation for thinking that narratives are necessary for model-based explanation is provided by any view of economic models that does not take them to establish causal mechanisms. For example, on the open formula view of models, models by themselves do not amount to a causal claim but only to a template for such a claim that needs to be filled in using knowledge from outside of the model (Alexandrova 2008). This view is motivated by the numerous idealizations in economic models that cannot be relaxed and cannot be part of a causal mechanism that explains the target phenomenon. Accordingly, a model must instead be treated as merely a template or open formula. It is only the causal claim developed on the basis of the open formula that does the explaining – and is what Morgan calls the narrative.

[5] Here is Ken Binmore doing such sidelining: "Such stories are not to be taken too seriously. Their chief purpose is to serve as a reminder about who gets what payoff" (Binmore 1994: 102).

[6] The revealed preference approach would reject this account of how Prisoner's Dilemma causally explains, denying that we need or should appeal to reasons. (Binmore himself holds this view, which might explain his dismissal of the role of narratives.) We discuss this in Section 4.4.2 below.

[7] This leaves open how situational analysis could be extended to cases where the actors in a Prisoner's Dilemma are not individual humans. We do not discuss that issue here, except to note we don't think there is any a priori reason why it couldn't be.

Accepting for now that this is how the Prisoner's Dilemma could explain, we move on to a potential obstacle. How could the Prisoner's Dilemma explain given that it is so *idealized*?

4.2.2 Prisoner's Dilemma and idealization

By the standards of other models in microeconomics, the Prisoner's Dilemma is remarkably undemanding. The simplest version can be formulated with only an ordinal utility function, not a cardinal one. As a result it needs only the minimal consistency axioms on preferences (completeness, transitivity, and asymmetry) and not the more controversial rankings of lotteries that the von Neumann–Morgenstern expected utility maximization framework requires. In addition to this the single-shot equilibrium, i.e. defection by both players, can be justified by dominance analysis only. It is thus not necessary to assume that players follow Nash equilibrium and hence co-ordinate their beliefs about each other. In this sense the Prisoner's Dilemma relies on far fewer controversial assumptions than do other models in game theory.

But it is still idealized nevertheless. It postulates an invariable choice of a dominant strategy by perfectly consistent agents. Actual people are not like this, as many experiments show, and that is already enough to query how such a model (or model plus narrative) could be explanatory. Can idealization ever be reconciled with explanation? Most certainly it can. Philosophers of science have come up with various accounts to make sense of the widespread explanatory use of seemingly false models.[8] We do not need to go into the details here. Roughly, they all come down to the same verdict: idealized models can be explanatory in the causal sense when their falsity does not matter, i.e. when the idealizations are true enough for the purposes at hand.

But for the Prisoner's Dilemma this defense will generally not work. Evidence from behavioral economics about how deeply context affects judgment and choice is robust. And social situations that approximate the single-shot or iterated Prisoner's Dilemma either in the field or in the laboratory exhibit a great deal of variability in levels of co-operation, enough to raise questions about the Prisoner's Dilemma's predictive value. Nevertheless, this still leaves open the possibility that the Prisoner's Dilemma does work in a few important cases. We turn to that issue now.

[8] See, for instance, recent work by Nancy Cartwright, Daniel Hausman, Uskali Mäki, Michael Strevens, and Michael Weisberg. For an overview, see Weisberg (2012).

4.3 The reality of non-explanation

4.3.1 Casual empiricism

Various encyclopedia entries and overview articles across economics and philosophy discuss some of the Prisoner's Dilemma literature's main developments: asymmetric versions, versions with multiple moves or players, single-person interpretations, versions with asynchronous moves, finitely and infinitely and indefinitely iterated versions, iterated versions with error, evolutionary versions, versions interpreted spatially, and many other tweaks besides (Govindan and Wilson 2008, Michihiro 2008, Kuhn 2009). Many of these are apparently motivated by a loose kind of responsiveness to real-world problems and complications. After all, putative actual players of the Prisoner's Dilemmas will often act asynchronously or more than once, or make errors, and so on. Certainly, the subtlety and sophistication of this work is often impressive. Nevertheless, a striking fact about it is its overwhelmingly *theoretical* focus. The underlying motivation by real-world concerns is somewhat casual. Deeper empirical applications of the Prisoner's Dilemma, featuring detailed examination of the evidence of particular real-world cases, are remarkably thin on the ground.

The overall picture is that research muscle has been bet on theoretical development rather than empirical applications.[9] It is in fact hard to find serious attempts at applying the Prisoner's Dilemma to explain actual historical or contemporary phenomena. We have found that the instances in which the Prisoner's Dilemma is mentioned in empirical contexts tend to come in two kinds. The first kind are the purely informal mentions in textbooks, blog posts, teaching tools, or offhand remarks in the media of the sort: "Well, that's obviously a Prisoner's Dilemma!"[10] Clearly, merely identifying a casual similarity between the Prisoner's Dilemma and an actual situation does not count as explanatory success. Sure, the price war between two gas stations may look like a Prisoner's Dilemma in some respects, but in other respects it doesn't. It would need to be explained why the dissimilarities do not matter.

The second kind of empirical use is far from casual. Ever since the discovery and proliferation of game theory in Cold War US academia, a great many fields in social science have adopted the language of the Prisoner's Dilemma (among other models) to reconceive old explananda, be they in industrial

[9] A lot of the Prisoner's Dilemma literature is "empirical" in the sense that it reports on psychological experiments. We discuss these in Section 4.4.1 below.

[10] http://cheaptalk.org/2013/11/13/prisoners-dilemma-everywhere-amazon-source/

organization or international bargaining (Erickson et al. 2013, Jervis 1978). But again, only rarely are game theory models applied carefully to specific field phenomena, and when they are it is not the Prisoner's Dilemma that is used. For the most part, the game theory models instead play a research-structuring rather than explanatory role, defining an agenda for the disciplines in question (see also Section 4.4.3).

4.3.2 A case study: Prisoner's Dilemma and World War I truces

Surveying the social sciences one finds a great many instances where the Prisoner's Dilemma is mentioned as explaining a field phenomenon. But the closer one looks, the more elusive explanatory success becomes. In the limited space here, we will support this claim via an extended analysis of one example. Of course, a single case does not prove much by itself. But if the Prisoner's Dilemma's explanatory shortcomings only become apparent when one looks at the fine details, then it is much more instructive to look at one case in depth than at many cases superficially.

The particular case we will examine is the "live-and-let-live" system that arose in World War I (WWI) trenches, which Robert Axelrod analyzed in terms of the Prisoner's Dilemma in chapter 4 of his book (1984). It is the most famous example of a detailed application of the Prisoner's Dilemma to a particular real-world target. It is also arguably the best one too, even though the details of Axelrod's analysis have subsequently been challenged (see Section 4.3.3 below).

Axelrod draws on the fascinating and detailed account of WWI trench warfare by the historian John Ashworth (1980), itself based on extensive letters, archives, and interviews with veterans. The "live-and-let-live" system refers to the many informal truces that arose on the Western front. "Truces" here covers complete non-aggression, temporary periods of non-aggression (e.g. at mealtimes), certain areas of non-aggression (e.g. mutually recognized "safe areas"), or many other mutual limitations on aggression (e.g. intricate local norms covering what actions and responses were or were not "acceptable"). The striking fact is that such truces between enemies arose spontaneously despite constant severe pressure against them from senior commanders. How could this have happened?

Axelrod's case is that, upon analysis, the implicit payoffs for each side on the front formed a Prisoner's Dilemma, and that this is an excellent example of how the Prisoner's Dilemma can illuminate a real-world phenomenon. In particular, he argues that the situation was an indefinitely iterated Prisoner's

Dilemma, and that co-operation – i.e. a truce – was therefore exactly the Prisoner's Dilemma's prediction.[11]

Axelrod is quite explicit that his goal is explanation, and of multiple explananda (1984: 71):

"The main goal [of the WWI case study] is to use the theory to explain:

1) How could the live-and-let-live system have gotten started?
2) How was it sustained?
3) Why did it break down toward the end of the war?
4) Why was it characteristic of trench warfare in World War I, but of few other wars?

A second goal is to use the historical case to suggest how the original concepts and theory can be further elaborated."

Of course, he is well aware of the many real-life complications. But he defends the application of the Prisoner's Dilemma nevertheless (1984: 19): "The value of an analysis without [the real-life complications] is that it can help to clarify some of the subtle features . . . which might otherwise be lost in the maze of complexities of the highly particular circumstances in which choices must actually be made. It is the very complexity of reality which makes the analysis of an abstract interaction so helpful as an aid to understanding."

Axelrod's meaning is less clear here, but perhaps his aims can be interpreted as some combination of explanation, heuristic value, and understanding, and maybe also the unificatory virtue of generalizability across contexts. Certainly, these seem very reasonable goals. Indeed, if applying the Prisoner's Dilemma did not achieve any of these, what would be the gain from applying it at all? So let us examine how well Axelrod's study fares by these criteria.

Many historical details do seem to tell in its favor:

- Breaches of a truce were followed by retaliation – but only on a limited scale. This is consistent with Tit-for-Tat.

[11] In fact, of course, the indefinitely iterated Prisoner's Dilemma has many other Nash equilibria besides mutual cooperation. The analysis that Axelrod actually applies comes from his well-known Prisoner's Dilemma computer tournaments, the winner of which he concluded was the Tit-for-Tat strategy with initial cooperation (Section 4.3.3). If adopted by both players, this strategy predicts indefinite mutual co-operation. Throughout this section, we will use "Prisoner's Dilemma" as shorthand for this richer theoretical analysis of Axelrod's. (The main lesson, namely the difficulty of establishing the Prisoner's Dilemma's explanatory success, would apply still more strongly to the Prisoner's Dilemma alone, because then we would be faced with the additional problem of equilibrium selection too.)

- Both sides often demonstrated their force capability – but in harmless ways, such as by expertly shooting up a harmless barn. Axelrod argues that Tit-for-Tat predicts that a credible threat is important to making co-operation optimal, but that actually defecting is not optimal. Hence, ways of establishing credibility in a non-harmful manner are to be expected.
- The Prisoner's Dilemma predicts that iteration is crucial to maintaining a truce. Soldiers actively sought to ensure the required continuity on each side, even though individual units were often rotated. For instance, old hands typically instructed newcomers carefully as to the details of the local truce's norms, so that those norms often greatly outlasted the time any individual soldier spent on that front.

Perhaps Axelrod's most striking evidence is how the live-and-let-live system eventually broke down. The (unknowing) cause of this, he argues, was the beginning of a policy, dictated by senior command, of frequent *raids*. These were carefully prepared attacks on enemy trenches. If successful, prisoners would be taken; if not, casualties would be proof of the attempt. As Axelrod observes:

> There was no effective way to pretend that a raid had been undertaken when it had not. And there was no effective way to co-operate with the enemy in a raid because neither live soldiers nor dead bodies could be exchanged. The live-and-let-system could not cope with the disruption . . . since raids could be ordered and monitored from headquarters, the magnitude of the retaliatory raid could also be controlled, preventing a dampening of the process. The battalions were forced to mount real attacks on the enemy, the retaliation was undampened, and the process echoed out of control. (Axelrod 1984: 82)

The conditions that the Prisoner's Dilemma predicts as necessary for co-operation were unavoidably disrupted and, Axelrod argues, it is no coincidence that exactly then the truces disappeared.

We agree that many of the historical details are indeed, in Axelrod's phrase, "consistent with" the situation being an iterated Prisoner's Dilemma.[12] Nevertheless, upon closer inspection, we do not think the case yields any predictive

[12] As we will see, many other details were *not* so consistent. But even if they all had been, this criterion is far too weak for explanation. After all, presumably the WW1 details are all consistent with the law of gravity too, but that does not render gravity explanatory of them.

or explanatory vindication of the Prisoner's Dilemma, contrary both to Axelrod's account and to how that account has been widely reported.

Why this negative verdict? To begin, by Axelrod's own admission some elements of the story deviate from his Prisoner's Dilemma predictions. First, the norms of most truces were not Tit-for-Tat but more like Three-Tits-for-Tat. That is, retaliation for the breach of a truce was typically three times stronger than the original breach.[13] Second, in practice two vital elements to sustaining the truces were the development of what Axelrod terms ethics and rituals: local truce norms became ritualized, and their observance quickly acquired a moral tinge in the eyes of soldiers. Both of these developments made truces much more robust and are crucial to explaining those truces' persistence, as Axelrod concedes. Yet, as Axelrod also concedes, the Prisoner's Dilemma says nothing about either. Indeed, he comments (1984: 85) that this emergence of ethics would most easily be modeled game-theoretically as a change in the players' payoffs, i.e. potentially as a different game altogether.

Moreover, there are several other predictive shortfalls in addition to those remarked by Axelrod. First, Tit-for-Tat predicts that there should be no truce-breaches at all. Again, this prediction is incorrect: breaches were common. Second, as a result (and as Axelrod acknowledges), a series of dampening mechanisms therefore had to be developed in order to defuse post-breach cycles of retaliation. Again, the Tit-for-Tat analysis is silent about this vital element for sustaining the truces. Third, it is not just that truces had to be robust against continuous minor breaches; the bigger story is that often no truces arose at all. Indeed, Ashworth examined regimental and other archives in some detail to arrive at the estimate that, overall, truces existed about one-quarter of the time (1980: 171–175). That is, on average, three-quarters of the front was *not* in a condition of live-and-let-live. Again, the Prisoner's Dilemma is utterly silent as to why. Yet part of explaining why there were truces is surely also an account of the difference from those cases where there were *not* truces.[14]

Moreover again, the Prisoner's Dilemma does not fully address two other, related issues. The first is how truces originated as opposed to how they

[13] The Prisoner's Dilemma itself (as opposed to Tit-for-Tat) is silent about the expected level of retaliation, so should stand accused here merely of omission rather than error.

[14] Ashworth, by contrast, does develop a detailed explanation, largely in terms of the distinction between elite and non-elite units, and their evolving roles in the war. The escalation in the use of raids, so emphasized by Axelrod, is only one part of this wider story. Most areas of the front were not in a state of truce even before this escalation.

persisted, about which it is again completely silent.[15] The second is how truces ended. This the Prisoner's Dilemma does partly address, via Axelrod's discussion of raids. But many truces broke down for other reasons too. Ashworth devotes most of his chapter 7 to a discussion of the intra-army dynamics, especially between frontline and other troops, which were often the underlying cause of these breakdowns.

And moreover once more, Ashworth analyses several examples of strategic sophistication that were important to the maintenance of truces but that are not mentioned by Axelrod. One such example is the use by infantry of gunners. In particular, gunners were persuaded to shell opposing infantry in response to opponents' shelling, so that opposing infantry would then pressurize their own gunners to stop. This was a more effective tactic for reducing opponents' shelling than any direct attack on hard-to-reach opposing gunners (168). Another example: the details of how increased tunnelling beneath enemy trenches also disrupted truces, quite separately from increased raiding (199–202). Perhaps Axelrod's analysis could be extended to these other phenomena too; but in lieu of that, the Prisoner's Dilemma's explanatory reach here seems limited.

We have not yet even mentioned more traditional worries about rational choice explanations. An obvious one here is that the explanations are after-the-fact; there are no novel predictions. Thus it is difficult to rule out wishful after-the-fact rationalization, or that other game structures might fit the evidence just as well. A second worry is that Axelrod's crucial arguments that the payoff structure fits that of an iterated Prisoner's Dilemma are rather brief and informal (1984: 75). Do his estimations here really convince?[16] And are the other assumptions of the Prisoner's Dilemma, such as perfectly rational players and perfect information, satisfied sufficiently well?

In light of these multiple shortfalls, how can it be claimed that the Prisoner's Dilemma explains the WWI truces? It is not empirically adequate, and it misses crucial elements even in those areas where at face value it is empirically adequate. Moreover, it is silent on obvious related explananda, some of them cited as targets by Axelrod himself: not just why truces persisted but also why they occurred on some occasions but not on others, how they originated, and (to some degree) when and why they broke down.

[15] Again, Ashworth covers this in detail (as Axelrod does report).

[16] Gowa (1986) and Gelman (2008), for instance, argue that they do not. (Gowa also voices some of our concerns about the explanatory adequacy of Axelrod's analysis as compared to Ashworth's.)

But note that there is no mystery as to what the actual causal explanations of these various explananda are, for they are given clearly by Ashworth and indeed in many cases are explicit in the letters of the original soldiers. Thus, for instance, elite and non-elite units had different attitudes and incentives, for various well-understood reasons. These in turn led to truces occurring overwhelmingly only between non-elite units, again for well-understood reasons. The basic logic of reciprocity that the Prisoner's Dilemma focuses on, meanwhile, is ubiquitously taken by both Ashworth and the original soldiers to be so obvious as to be mentioned only briefly or else simply assumed. Next, why did breaches of truces occur frequently, even before raiding became widespread? Ashworth explains via detailed reference to different incentives for different units (artillery versus frontline infantry, for instance), and to the fallibility of the mechanisms in place for controlling individual hotheads (1980: 153–171). And so on. Removing our Prisoner's Dilemma lens, we see that we have perfectly adequate explanations already.

Overall, we therefore judge both that the Prisoner's Dilemma does not explain the WWI truces, and that we already have an alternative – namely, historical analysis – that does. So if not explanation, what else might the Prisoner's Dilemma offer? What fallback options are available? It seems to us there are two. The first is that, explanatory failure notwithstanding, the Prisoner's Dilemma nevertheless does provide a deeper "insight" or "understanding," at least into the specific issue of why the logic of reciprocity sustains truces. We address this response elsewhere (Northcott and Alexandrova 2013). In brief, we argue that such insight is of no independent value without explanation, except perhaps for heuristic purposes.

This leads to the second fallback position – that even if the Prisoner's Dilemma does not provide explanations here, still it is of heuristic value (see also Section 4.4.3 below). In particular, presumably, it is claimed to guide us to those strategic elements that do provide explanation. So does the Prisoner's Dilemma indeed add value in this way to our analysis of the WWI truces? Alas, the details suggest not, for two reasons.

First, the Prisoner's Dilemma did not lead to any causal explanations that we didn't have already. To see this, one must note a curious dialectical ju-jitsu here. Axelrod cites many examples of soldiers' words and actions that seem to illustrate them thinking and acting in Prisoner's Dilemma-like patterns. These are used to support the claim that the Prisoner's Dilemma is explanatory. (This is a common move in casual applications of the Prisoner's Dilemma more generally.) Yet now, having abandoned the explanatory claim and considering instead whether the Prisoner's Dilemma might be valuable heuristically, these very same examples become evidence *against* its value rather than for it. This is because they now show that Prisoner's Dilemma-like thinking was present already. Ubiquitous quotations in Ashworth, many cited by Axelrod himself, show that soldiers were very well aware of the basic strategic logic of reciprocity. They were also well aware of the importance of a credible threat for deterring breaches (Ashworth 1980, 150). And well aware too of why frequent raiding rendered truces impossible to sustain, an outcome indeed that many ruefully anticipated even before the policy was implemented (Ashworth 1980: 191–198).[17]

The second reason why the Prisoner's Dilemma lacks heuristic value is that it actively diverts attention *away* from aspects that are important. We have in mind many of the crucial features already mentioned: how truces originated, the causes and management of the continuous small breaches of them, the importance of ethics and ritualization to their maintenance independent of strategic considerations, why truces occurred in some sections of the front but not in a majority of them, and so on.[18] Understanding exactly these features is crucial if our aim is to encourage co-operation in other contexts too – and this wider aim is the headline one of Axelrod's book,[19] and implicitly surely a

[17] Ashworth reports (1980: 197): "One trench fighter wrote a short tale where special circumstances … [enabled the truce system to survive raids]. The story starts with British and Germans living in peace, when the British high command wants a corpse or prisoners for identification and orders a raid. The British soldiers are dismayed and one visits the Germans taking a pet German dog, which had strayed into British trenches. He attempts to persuade a German to volunteer as a prisoner, offering money and dog in exchange. The Germans naturally refuse; but they appreciate the common predicament, and propose that if the British call off the raid, they could have the newly dead body of a German soldier, providing he would be given a decent burial. The exchange was concluded; the raid officially occurred; high command got the body; and all parties were satisfied. All this is fiction, however…"

This soldier's fictional tale demonstrates vividly a very clear understanding of the Prisoner's Dilemma's strategic insights *avant la lettre*, indeed a rather more nuanced and detailed understanding than the Prisoner's Dilemma's own. No need for heuristic aid here.

[18] For example, a full understanding of why raiding disrupted truces goes beyond the simple Prisoner's Dilemma story. Ashworth summarises (1980: 198): "Raiding … replaced a background expectancy of trust with one of mistrust, making problematic the communication of peace motives; raids could not be ritualised; the nature of raids precluded any basis for exchange among adversaries; and raiding mobilised aggression otherwise controlled by informal cliques."

[19] Axelrod summarizes (1984: 21–22) the wider lessons of the WWI case for cooperation in this way: it can emerge spontaneously, even in the face of official disapproval; it can be tacit rather than explicit; it requires iterated interaction; and it does not require friendship between the two parties. But all these lessons are already contained in Ashworth's historical account – and, we argue, Ashworth establishes them rather better.

major motivation for the Prisoner's Dilemma literature as a whole. Yet here, to repeat, the Prisoner's Dilemma directs our attention away from them!

Overall, in the WWI case:

1) The Prisoner's Dilemma is not explanatory.
2) The Prisoner's Dilemma is not even valuable heuristically. Rather, detailed historical research offered much greater heuristic value, as well as much greater explanatory value.

Thus, Axelrod's own stated goals were not achieved. More generally, if this case is indicative then we should conclude that, at least if our currency is causal explanations and predictions of real-world phenomena, the huge intellectual investment in the Prisoner's Dilemma has not been justified.

4.3.3 It's not just Axelrod

Axelrod's work was innovative in that he arrived at his endorsement of Tit-for-Tat via a simulation rather than by calculation. For this reason, he has been credited with helping to kick-start the research program of evolutionary game theory. His engaging presentation also quickly won a popular following. Nevertheless, even theorists sympathetic to the potential of the Prisoner's Dilemma to explain cooperation have since then largely rejected the details of his analysis – not on the empirical grounds that we have emphasized, but rather on theoretical grounds. In particular, other simulations have not reproduced Tit-for-Tat's superiority; indeed, often "nasty" strategies are favored instead (e.g. Linster 1992). More generally, Axelrod's approach arguably suffers badly from a lack of connection to mainstream evolutionary game theory (Binmore 2001). The conclusion is that it is dubious that the WWI soldiers should be predicted to play Tit-for-Tat at all.

It does not follow, however, that Axelrod is therefore a misleadingly easy target – for two reasons. First, no better analysis of the WWI case has appeared. What strategy does best model soldiers' behavior in the trenches? This is neither known, nor has anyone bothered to find out. It is true that there are now much more sophisticated results from simulations of iterated Prisoner's Dilemma in different environments and, thus, better theoretical foundations. But there has been no attempt to use these improved foundations to model the WWI live-and-let-live system. Until a successor analysis has actually been applied to the WWI case, we have no reason to think it would explain the behavior in the trenches any better than did Axelrod's, let alone better than Ashworth does.

Second, it is not just that the WWI case in particular has been left ignored by the emphasis on theory. Rather, it is that the same is true of field cases generally. Detailed empirical engagement is very rare.[20] Of course, short of an exhaustive survey it is hard to prove a negative thesis such as this, but we do not think the thesis is implausible. One initial piece of evidence is that Axelrod's WWI study continues to be used in many textbooks as a prime example of the Prisoner's Dilemma's supposed explanatory relevance.[21] Perhaps these textbooks' selections are just ill judged, but the point is the perceived lack of alternative candidates.

Or consider the career of the Prisoner's Dilemma in biology – a discipline often cited by game theorists as fertile ground for applications. But the details turn out to be discouraging there too, and for a familiar reason, namely a focus on theoretical development rather than on field investigations:

> [T]he preoccupation with new and improved strategies has sometimes distracted from the main point: explaining animal cooperation . . . Understanding the ambiguities surrounding the Iterated Prisoner's Dilemma has stimulated 14 years of ingenious biological theorizing. Yet despite this display of theoretical competence, there is no empirical evidence of non-kin cooperation in a situation, natural or contrived, where the payoffs are known to conform to a Prisoner's Dilemma. (Clements and Stephens 1995)

And for a similarly negative verdict:

> [D]espite the voluminous literature, examples of Prisoner's Dilemma in nature are virtually non-existent. . . Certainly, with all the intense research and enthusiastic application of [the Prisoner's Dilemma] to real world situations, we may expect that we should have observed more convincing empirical support by now if it ever were to hold as a paradigm. . . (Johnson et al. 2002)

Payoff structures in field cases rarely seem to match those of the Prisoner's Dilemma, often because of the different values put on a given outcome by different players. Johnson et al. (2002) explain why several much-reported successes are in fact only dubiously cases of Prisoner's Dilemma at all, such as predator "inspection" in shoaling fish, animals cooperating to remove

[20] Sunstein (2007) comes close, but even here the phenomenon in question (the failure of the Kyoto Protocol) is explained in part by the fact that it does *not* have a Prisoner's Dilemma structure.

[21] E.g. Besanko and Braeutigam (2010: 587–588) – and there are many other examples.

parasites from each other, or lions cooperating to defend territory. The one exception they allow is Turner and Chao's (1999) study of an RNA virus. Even the game theorists Nowak and Sigmund (1999: 367), while lionizing the Turner and Chao case, concede that other claimed cases of the Prisoner's Dilemma occurring in nature are unproven. They also concede that, with reference to the literature in general, "it proved much easier to do [computer] simulations, and the empirical evidence lagged sadly behind."

Nor does there seem good reason to expect a dramatically different story in other disciplines. Gowa (1986), for instance, in a review of Axelrod's 1984 book, is generally sympathetic to the application of formal modeling. Nevertheless, she argues that the simple Prisoner's Dilemma template is unlikely to be a useful tool for studying the complex reality of international relations. And indeed since then bargaining models have become the norm in IR, because they can be purpose-built to model specific cases of negotiations in a way that the Prisoner's Dilemma can't be (e.g. Schultz 2001).

Overall, the Axelrod WWI case is therefore not a misleadingly soft target amid a sea of many tougher ones. On the contrary, it remains by far the most famous detailed application of the Prisoner's Dilemma to a field case for good reason – there aren't many others.

4.4 Defenses of Prisoner's Dilemma

4.4.1 Laboratory experiments

As we have noted, a large portion of the Prisoner's Dilemma literature concerns theoretical development, in which we include the running of the dynamics of idealized systems. Very little concerns close empirical analysis of field phenomena. But there is a third category that, although it is hard to quantify precisely, in terms of sheer number of papers might form the largest portion of all. This third category concerns psychology experiments, in particular simulation in the laboratory of the Prisoner's Dilemma or closely related strategic situations. Do the human subjects' actions in the laboratory accord with the predictions of theory? What factors are those actions sensitive to? Even a cursory sampling of the literature quickly reveals many candidates. For example, how much is cooperation in a laboratory setting made more likely if we use labeling cues (Zhong et al. 2007), if we vary payoffs asymmetrically (Ahn et al. 2007), if there is a prior friendship between players (Majolo et al. 2006), if players have an empathetic personality type (Sautter et al. 2007), or if players expect cooperation from opponents (Acevedo and Krueger

2005)? Literally thousands of articles are similar. Do they demonstrate, as it were, an empirical wing to the Prisoner's Dilemma literature after all? Unfortunately we think not, or at least not in the right way. Here are two reasons for this negative verdict.

First, the emphasis in most of this literature is on how a formal Prisoner's Dilemma analysis needs to be *supplemented*.[22] Typically, what makes cooperation more likely is investigated by manipulating things external to the Prisoner's Dilemma itself, such as the psychological and social factors mentioned above. That is, the focus of the literature is on how the Prisoner's Dilemma's predictions break down and on how instead a richer account, sensitive to otherwise unmodeled contextual factors, is necessary to improve predictive success. This is just the same lesson as from the WWI case – only now this lesson also holds good even in the highly controlled confines of the psychology laboratory.

Second, an entirely different worry is perhaps even more significant: whatever the Prisoner's Dilemma's success or otherwise in the laboratory, what ultimately matters most is its success with respect to *field* phenomena. Does it predict or explain the behavior of banks, firms, consumers, and soldiers outside the laboratory? Surely, that must be the main motivation for social scientists to use the Prisoner's Dilemma. Accordingly, the main value of the psychology findings, at least for non-psychologists, must be *instrumental* – are they useful guides to field situations? Presumably, they would indeed be if the psychological patterns revealed in experiments carried over reliably to field cases. Suffice to say here that such extrapolation is far from automatic, given the huge range of new contextual cues and inputs to be expected whenever moving from the laboratory to the field. The issue is the classic one of external validity, on which there is a large literature.[23] So far, the field evidence for the Prisoner's Dilemma is not encouraging.

[22] As Binmore and Shaked (2010) and others argue, other empirical work shows that, after a period of learning, the great majority of laboratory subjects do eventually defect in one-shot Prisoner's Dilemma games, just as theory predicts. Nevertheless it is uncontroversial that, initially at least, many or even most do not. It is this that has spawned the large literature investigating what contextual factors influence such instances of cooperation.

[23] Levitt and List (2007) discuss this from an economist's perspective with regard to cooperation specifically. Like everyone else, they conclude that external validity can rarely if ever be assumed. This is true even of field explananda that one might think especially close to laboratory conditions and thus especially promising candidates, such as highly rule-confined situations in TV game shows (see, e.g., van den Assem et al. 2012 about the Split or Steal show).

4.4.2 Revealed preferences to the rescue?

There is another way to defend the Prisoner's Dilemma's explanatory power. According to it, the Prisoner's Dilemma is not supposed to furnish explanations in which people co-operate because they *feel* it would be better for them and they can *trust* the other party to reciprocate; or fail to cooperate because they are *afraid* of being taken for a ride. Although these are the conventional articulations of what happens in a Prisoner's Dilemma, they are causal claims made using psychological categories such as feelings, judgments, and fears. They assume that behavior stems in part from these inner mental states and can be explained by them.

But a long tradition in economics maintains that this is exactly the wrong way to read rational choice models. Agents in these models do not make choices because they judge them to be rational; rather, the models are not psychological at all. To have a preference for one option over another just *is* to choose the one option when the other is available. This is the well-known revealed preference framework. It *defines* preferences as choices (or hypothetical choices), thus enforcing that economic models be interpreted purely as models that relate observable behavior to (some) observable characteristics of social situations.[24] On this view, agents cooperate in the Prisoner's Dilemma not because they feel they can trust each other, but rather because this is a game with an indefinite horizon in which the payoffs are such that rational agents cooperate. Although such an explanation sounds tautologous, it isn't. It comes with relevant counterfactual claims, such as that (given their history of choices) agents would not have cooperated if the game had been single-shot rather than iterated. This is a causal counterfactual and thus can be used for causal explanation. It only sounds tautologous because we are used to the natural and deeper psychological reading of the Prisoner's Dilemma in line with standard explanations of actions. But the revealed preference reading is perfectly conceivable too, and moreover the party line in economics is that it is in fact the correct one.

We will not discuss why the revealed preference view became popular within economics, nor evaluate whether it is viable in general.[25] Rather, our interest here is whether even according to it the Prisoner's Dilemma is a promising research program for explaining actual field cases. On this latter issue, we make two pessimistic points.

[24] Only "some" because, on the revealed preference view, data on what agents say, or on their physiological and neurological properties, are typically not deemed admissible even though they are perfectly observable.

[25] For up-to-date interpretations, criticisms, defenses, and references, see Hausman (2012).

First, a strict revealed preference theory of explanation seems needlessly philistine. To the extent that we have a good explanation for the live-and-let-live system in the WWI trenches it is in part a psychological explanation deeply steeped in categories such as fear, credibility, and trust. This is a general feature of social explanations – they are explanations that appeal to beliefs and desires (Elster 2007). For the revealed preference theorist, this is reason to dump them. But Ashworth's WWI explanations would be greatly impoverished if we did. In fact, not much of his rich and masterful analysis would remain at all.

Second, even if interpreted in revealed preference terms, the Prisoner's Dilemma would still state false counterfactual (or actual) claims. Many more factors affect behavior than just the ones captured by the Prisoner's Dilemma. But the revealed preference defense only works if an explanation is empirically adequate (ignoring for now its false behavioral claims about how people reason). And the Prisoner's Dilemma's explanations aren't empirically adequate even in the very cases that are deemed to be its great successes, or so we have argued. In which case, the revealed preference defense fails.

4.4.3 An agenda setter?

Even if the Prisoner's Dilemma does not explain many social phenomena, might it still play other useful roles? We will discuss here two candidates. The first role, mentioned earlier, is *heuristic*. More particularly, the thought is that even if it were not directly explanatory of individual cases, still the Prisoner's Dilemma might serve as an agenda-setter, structuring research. Descriptively speaking, there is much reason to think that this has indeed happened. But normatively speaking, is that desirable? Maybe sometimes. For example, from the beginning the Prisoner's Dilemma was lauded for making so clear how individual and social optimality can diverge. Moreover, it seems convincing that it has been heuristically useful in some individual cases, such as in inspiring frameworks that better explain entrepreneur–venture capitalist relations (Cable and Shane 1997). This would replicate the similar heuristic value that has been claimed for rational choice models elsewhere, for instance in the design of spectrum auctions (Alexandrova 2008, Northcott and Alexandrova 2009).

Nevertheless, overall we think there is reason for much caution. At a micro level, it is all too easy via casual empiricism to claim heuristic value for the Prisoner's Dilemma when in fact there is none. The WWI example illustrates this danger well – there, the Prisoner's Dilemma arguably turned out to be of *negative* heuristic value. On a larger scale, we have seen the gross disproportion between on one hand the huge size of the Prisoner's Dilemma literature and on the other hand the apparently meager number of explanations of field phenomena that this literature has achieved. Overall, the concentration on theoretical development and laboratory experiments has arguably been a dubious use of intellectual resources.

4.4.4 A normative role?

The second non-explanatory role that the Prisoner's Dilemma might serve is to reveal what is instrumentally rational. Even if it fails to predict what agents actually did, the thought runs, still it might tell us what they *should* have done. For example, given their preferences, two battalions facing each other across WWI trenches would be well advised to cooperate; that is, if the situation is such that they face an indefinitely repeated Prisoner's Dilemma, then it is rational not to defect.

There is an obvious caveat to this defense though, explicit already in its formulation: the normative advice is good only if the situation is indeed accurately described as a Prisoner's Dilemma. Thus a normative perspective offers no escape from the central problem, namely the ubiquitous significance in practice of richer contextual factors unmodeled by the Prisoner's Dilemma.

4.4.5 The aims of science

Why, it might be objected, should the goal of social science be mere causal explanations of particular events? Isn't such an attitude more the province of the historian? Social science should instead be concentrating on systematic knowledge. The Prisoner's Dilemma, this objection concludes, is a laudable example of exactly that – a piece of theory that sheds light over many different cases.

In reply, we certainly agree that regularities or models that explain or that give heuristic value over many different cases are highly desirable. But ones that do neither are not – especially if they use up huge resources along the way. When looking at the details, the Prisoner's Dilemma's explanatory record so far is poor and its heuristic record mixed at best. The only way to get a reliable sense of what theoretical input would actually be useful is via detailed empirical investigations. What useful contribution – whether explanatory, heuristic, or none at all – the Prisoner's Dilemma makes to such investigations cannot be known until they are tried. Therefore resources would be better directed towards that rather than towards yet more theoretical development or laboratory experiments.

5 The Prisoner's Dilemma and the coevolution of descriptive and predictive dispositions

Jeffrey A. Barrett

5.1 Introduction

In the standard Prisoner's Dilemma there are two players *A* and *B*. Each player has the option of cooperating or defecting in a single-shot play of the game; each decides what to do without interacting with the other, and their decisions are revealed simultaneously. Suppose that if both players choose to cooperate, they each spend *one year* in prison; if both choose to defect, they each spend *two years* in prison; and if one chooses to cooperate and the other to defect, the defector *goes free* and the cooperator spends *three years* in prison. Finally, suppose that each player is ideally rational and has perfect and complete knowledge concerning the precise nature of the game being played.

Player *A* reasons as follows. Player *B* might cooperate or defect. If *B* cooperates, then I would do better by defecting than by cooperating. And if *B* defects, then I would do better by defecting than by cooperating. So regardless of what *B* does, I would do better by defecting. Player *B* reasons similarly. So both players defect. And since neither player can benefit by changing strategies when the other defects, mutual defection is a Nash equilibrium of the game, and the only such equilibrium.

The curious feature of the Prisoner's Dilemma is that by defecting each player does worse than had they both cooperated. Here, perfect rationality with perfect and complete knowledge leads to behavior that is suboptimal for both players, both individually and jointly.[1] Perfect rationality and perfect and complete knowledge are hence not necessarily virtues in the context of social interaction, at least not on this analysis.

I would like to thank Brian Skyrms, Jorge Pacheco, Thomas Barrett, and Justin Bruner for helpful discussions and comments.

[1] Indeed, on the standard assumption of perfect common knowledge, the agents know that they are doomed to act suboptimally before they even play.

Human agents, however, often cooperate in situations that at least look very like versions of the Prisoner's Dilemma.[2] There are a number of explanatory options at hand. An interaction that looks like a Prisoner's Dilemma might in fact be a very different game. Indeed, as we will see, it might be a game with features that are not well characterized by classical game theory at all. Further, it is clearly inappropriate to assume that human agents are even typically rational or well-informed. And here this may well serve both their individual and joint interests. Players who choose their strategies on the basis of similar fortune cookies – "Today is a good day to cooperate. Your lucky numbers are 2, 5, 7, 16, 33, 39." – rather than on the basis of rational deliberation with perfect knowledge, may end up cooperating to the benefit of each.

While one might readily grant that nonhuman agents (primates more generally, animals, insects, plants, single-cell organisms, and/or implemented algorithms) might not be ideally rational or perfectly well-informed, how human agents may also fall short of such idealizations is important. Particularly salient among our failings, there is good reason to believe that we often, perhaps even typically, lack reliable introspective access to both the cognitive processes by which we make decisions and the environmental factors that are in fact most salient to our choices.[3] When this happens, our decisions are manifestly not the result of rational deliberation in the context of complete and reliable knowledge and are hence not well characterized by any analysis that makes such idealizing assumptions.

A more broadly applicable approach to modeling decisions is to track the evolving dispositions of agents on repeated interactions. Such evolutionary games provide a way to consider how agents' dispositions might evolve regardless of how rational they may be or what, if anything, they may know about themselves, other agents, or the world they inhabit. Evolutionary games also allow one to consider how such constraints as who plays with whom and who has access to what information and when may affect the agents' evolving dispositions.

[2] Andreoni and Miller (1993) provide evidence for human subjects of both the cooperative effects of reputation repeated games and of robust baseline cooperative behavior in single-shot games.

[3] There is a large empirical literature on this aspect of human nature including Maier's (1931) early empirical studies on failures of introspection regarding problem solving, Nisbett and Wilson's (1977) classic survey paper on failures of introspection regarding cognitive processes, and Johansson et al.'s (2013) and Hall et al.'s (2010) recent experiments on failures of introspection concerning how we make decisions and justify our choices.

The strategy here will be to consider the Prisoner's Dilemma in a number of evolutionary contexts that each allow for the evolution of a variety of cooperation. The last of these is a version of the Prisoner's Dilemma with pre-play signaling, where the signals may influence whether the players cooperate or defect. On this model, their descriptive and strategic dispositions coevolve in an arms race where cooperators seek to identify each other through their evolving self-descriptions, and defectors seek to exploit these self-descriptions in order to play with cooperators. We will then consider a variety of Skyrms-Lewis signaling game that also exhibits coevolving descriptive and predictive dispositions. Finally, we will then briefly return to consider the Prisoner's Dilemma with pre-play signaling.

5.2 The Prisoner's Dilemma in an evolutionary context

It is customary to distinguish between two types of evolutionary game. In a population game, one tracks how a community of agents with various fixed strategies might evolve in terms of the proportion of each strategy represented in the population at a time. Here the dynamics of the game tells one how dispositions of current agents are passed to future generations on the basis of their success and failure. In contrast, in a learning game, one tracks how a fixed community of agents learn as they interact with each other in terms of how the dispositions of each agent evolves. Here, the dynamics of the game tells one how the agents update their dispositions to act in the future on the basis of their successes and failures. In each case, one is tracking the dispositions of agents.

Simply moving to an evolutionary context does not by itself explain why one might expect to see cooperation in the Prisoner's Dilemma. Consider, a population of cooperators and defectors who are repeatedly paired up at random to play a version of the standard Prisoner's Dilemma. Let $P_C(t)$ be the proportion of cooperators and $P_D(t)$ be the proportion of defectors in the population at time t. If the population is large, one might expect these proportions to evolve in a way determined by how the average fitness of cooperators $F_C(t)$ and the average fitness of the defectors $F_D(t)$ each compare to the average fitness of the total population $F(t)$. In particular, one might imagine that the proportion of a particular strategy type in the next generation $t + 1$ is determined by the current proportion of that type weighted by the ratio of the average fitness of that type against the average fitness of the entire population. Or more precisely, that $P_C(t + 1) = P_C(t)F_C(t)/F(t)$ and $P_D(t + 1) = P_D(t)F_D(t)/F(t)$. This expression is an example of the replicator dynamics. The idea is that one might expect strategies to be copied to the next generation in proportion to how successful they are in the current generation relative to the average success of the entire population.

Here, as in the classical Prisoner's Dilemma, cooperators do well when they play against other cooperators. Indeed, a population made up only of cooperators is an equilibrium of the replicator dynamics. But it is an unstable equilibrium. While cooperators do well playing with other cooperators, defectors do better than cooperators against both cooperators and other defectors. Defectors are, consequently, able to invade a population of cooperators. A mutant defector in a population of cooperators will do significantly better than any cooperator and hence be proportionally better represented in the next generation. Of course, as the proportion of defectors increases, they do not do as well as when they were first introduced into the population since they meet each other increasingly often, but, even then, they do better than a cooperator would. Cooperation, then, is driven to extinction.

A population made up of only defectors is also an equilibrium of the replicator dynamics. But, in contrast to pure cooperation, pure defection is a stable equilibrium. Cooperators cannot invade the population since they do worse in every interaction than the defectors. Indeed, since no mutant does better against defectors than defectors do against each other, defection is an evolutionarily stable strategy.[4] So while a pure population of cooperators does better overall than a pure population of defectors, the former is unstable and quickly evolves to the latter on the introduction of mutation.

While moving to an evolutionary context does not by itself provide an account of cooperation, it makes room for dynamical considerations that may. Axelrod and Hamilton (1981) show how agents may come to cooperate in order to protect their interests in possible future interactions in the context of type of learning game. Here, two agents play the Prisoner's Dilemma repeatedly, but they do not know beforehand how many times they will play. A strategy is a decision rule that determines the probability of cooperation or defection as a function of the history of the interactions between the agents so far. Each agent adopts a strategy to guide their play throughout. If the agents knew ahead of time how many times they would play, then defecting would be the only evolutionarily stable strategy since defection on the last play would clearly be optimal for each agent, and if they

[4] See Alexander (2014) for an extended discussion of the Prisoner's Dilemma in the context of replicator dynamics. See Taylor and Jonker (1978) for a discussion of the distinction between stability under game dynamics and the notion of an evolutionarily stable strategy.

both knew that they would defect on the last play, then they should also defect on the next to the last play, and so on.

But the situation is different if the number of plays is indefinite with a constant probability of a next play. Here, unconditional defection is a stable strategy, but there may also be other stable strategies if the probability of a next play is sufficiently high.[5] Rational agents may, for example, choose to play a grim-trigger strategy, where each cooperates until the other defects, or a Tit-for-Tat strategy, where each cooperates on their first play then does whatever the other did on the most recent play from then on. If both agents played either of these strategies, then they would cooperate on every play of the game and mutant defectors would be punished. Axelrod and Hamilton were also impressed by the fact that Tit-for-Tat does very well in simulation against much more sophisticated learning strategies.[6]

Other evolutionary mechanisms may also promote cooperation. Defectors easily invade a population of cooperators on the replicator dynamics because an individual defector does better than an individual cooperator against each type of player and because the players are matched up randomly. If cooperators only played with cooperators and defectors only played with defectors, then cooperation would flourish. But to simply stipulate anything like that would clearly be ad hoc. That said, it is clearly possible that it is, for one reason or another, more likely for a cooperator to play with a cooperator than with a defector in a particular real-world population.

One way this might happen is by dint of spacial proximity. Consider agents distributed on a lattice who can only play with neighboring agents at connected points on the lattice. An island of cooperators would only be exposed to defectors along its edges. Playing with each other in the center of the island, cooperators would flourish. Of course, the unfortunate cooperators along the edges would often play defectors, and those fortunate defectors would also flourish. Precisely what happens on such a model, then, depends on the

[5] A strategy might, for example, be neutrally stable in that a mutant does not do statistically better against the population.

[6] They report (Axelrod and Hamilton 1981, Axelrod 1984) that the Tit-for-Tat strategy won in a round-robin tournament of fourteen strategies. The more sophisticated strategies that it beat included one that modeled the observed behavior of the other agents as Markov processes, then used Bayesian inference to choose its own actions. Importantly, however, that a strategy is both simple and broadly successful does not entail that it is evolutionarily favored in all contexts. What strategies do well depends, in part, on what other strategies are in the population. More generally, often the best one can do is to provide a concrete model where the particular strategy typically evolves given the specific assumptions of the model.

payoffs of the particular version of the Prisoner's Dilemma played, the connectivity of the lattice, and the evolutionary dynamics one considers.

Nowak and May (1992, 1993) describe an example of such a game. Here, agents play a version of the Prisoner's Dilemma on a square lattice. The particular version of the Prisoner's Dilemma is specified by stipulating the payoffs for cooperating or defecting conditional on what the other agent does. Each node on the lattice is occupied by an agent. Each round, each agent plays the specified game with her eight nearest neighbors. At the end of each round, each agent compares her total payoff to that of each of these neighbors. She then adopts whatever strategy was used by the most successful of the nine agents, including herself, and plays that strategy in the next round.

For some settings of the payoffs, the population evolves to one where everyone defects. For other settings, the population evolves to one where both cooperators and defectors survive in either (1) a stable cycle with both cooperators and defectors at each step, or (2) a never-ending battle where defectors invade shifting islands of cooperators and cooperators invade shifting islands of defectors.[7]

Another way that cooperation might evolve would be if cooperators knew how to identify the strategy types of other agents and were able to opt out of play when paired with a defector.[8] This might happen if the likelihood of an agent cooperating or defecting on the next play of the game were correlated with some observable property of the agent. Perhaps players with blue hats are more likely to cooperate than players with red hats, or perhaps players who have been observed to cooperate in the past are more likely to cooperate in the future.

Rather than simply giving agents the ability to identify the strategies of other agents directly, one might consider agents who must learn how the observable features of other agents are in fact correlated to their dispositions. If a game is only played when neither player opts out, then cooperation might evolve if agents can learn to identify the strategy type of their perspective partner and then only play with likely cooperators and opt out with likely defectors. Even very simple reinforcement learners, like those we will consider in the next section, might readily discover such reputational correlations if they exist and are stable over time. If cooperators learn to opt out when they

[7] See Alexander (2014) for further details and beautiful illustrations of such evolutions.

[8] See Kitcher (1993) for a model of the evolutionary Prisoner's Dilemma that involves agents who can identify other agent types and opt out of play on the basis of the identification.

are paired up to play defectors and if play between cooperators is better than not playing at all, then cooperation can be expected to flourish.

Here, cooperation results from agents conditioning their play on learned regularities concerning the observable properties and strategic dispositions of other agents. But the ability to learn things like this sets up the possibility of an evolutionary arms race where defectors might also profit, and profit more, by imitating the property that is initially indicative of an increased likelihood of cooperation. Cooperation may still evolve, but the situation now is significantly more subtle.

Costless preplay signaling, also known as cheap-talk, furnishes an example of such an arms race. On Santos, Pacheco, and Skyrms' (2011) model of the Prisoner's Dilemma with pre-play signaling, agents are paired at random, each sends a signal, then they play a version of the Prisoner's Dilemma using strategies that may depend on the pre-play signals. In particular, the model allows that the pre-play signal an agent receives may come to determine the likelihood that she cooperates or defects when the game is played.[9]

To begin, suppose that the players' strategies do not depend on the signals they receive, and since they are entirely uncorrelated to their actions, their signals are meaningless. Now consider a mutant who always sends a particular, previously unused, signal and cooperates if and only if the other agent sends the same signal. This signal may then become a secret handshake between such agents. The signal will come to identify a conditional cooperator. Agents who use it will cooperate with each other whenever they meet, defect on other types of agents, and they will flourish.

But just as such a secret handshake is formed, it may be subverted. Consider a new type of mutant agent who sends the handshake but then defects regardless of the signal he receives. Such an agent will defect on agents who cooperate on the signal and will flourish in comparison. Unchecked, the signal will evolve to identify a sure-fire defector.

But this, in turn, sets up an opportunity akin to the first where a third type of mutant agent, one who uses a fresh signal type as a secret handshake, may flourish. Such agents will dominate both agents who are cooperating and agents who are defecting on old handshakes. And so on. The picture then is one where cooperators and defectors are locked in an evolving arms race

[9] This model builds on earlier work on pre-play signaling by Robson (1990) and follows Skyrms (2002). Allowing for the evolution of pre-play signaling here provides a model of what is sometimes referred to as the green-beard effect in biology (Hamilton 1964a, Hamilton 1964b).

where signals are used as secret handshakes allowing for cooperation, then secret handshakes are subverted leading to new secret handshakes.

While the agents' signaling behavior is constantly evolving, Santos, Pacheco, and Skyrms (2011) show that, for finite populations in the small mutation limit, this evolutionary process nevertheless allows for a stable cooperation. The degree of cooperation depends both on the size of the population and the number of signal types available. If there is only one possible signal type, the situation is identical to playing without signals and defection dominates. But if there are two signal types that might be used as the basis of a secret handshake, and a population of 150 agents for example, the agents will spend about 32 percent of their time cooperating in the limit.[10] And for a given population size, as the number of available signal types increases, the amount of time spent cooperating in the limit increases. The more available signal types there are, the harder it is for potential subverters to find the signal being used as a secret handshake in order to exploit it, and the better the cooperators do.

This model of the Prisoner's Dilemma with pre-play signaling tracks the coevolution of basic descriptive and strategic dispositions. More informally, it illustrates how it is possible for agents to coevolve a simple descriptive language and a type of predictive knowledge concerning how the descriptions of agents are correlated with their future behavior.

We will consider the coevolution of descriptive and predictive dispositions in the context of a very simple sort of common-interest signaling game and then briefly return to discuss the conflicting-interest Prisoner's Dilemma with pre-play signaling.

5.3 The coevolution of description and prediction

Sender–predictor games are a type of Skyrms–Lewis signaling game that provide basic models for the coevolution of simple descriptive and predictive dispositions.[11] We will consider a sequence of three such games. The first is a

[10] The population spends most of the time in one monomorphism or another.

[11] For discussions of basic Skyrms–Lewis signaling games and the more elaborate sender–predictor games, see Lewis (1969), Skyrms (2006), Argiento et al. (2009), Barrett (2007), Barrett (2009), and Skyrms (2010). Insofar as the receiver might always be taken to be performing a predictive action based on a prior state observed by the sender, there is a sense in which a basic Skyrms–Lewis signaling game is also a sender–predictor game with deterministic laws relating the prior and posterior states of nature. See Barrett (2007) and Barrett (2009) for further details regarding the games discussed in this section.

simple two-state, one-sender game, the second is a two-sender, four-state game, and the third is a two-sender, four-state game with payoffs that change during play. We will treat these evolutionary games in a learning context.

In the simplest type of sender–predictor game there are two agents: a sender and a predictor. The sender observes the prior state of nature, then sends a signal. The predictor, who does not have direct access to the state of nature, receives the signal and then performs a predictive action that either matches the posterior state of nature at some later time and is successful or does not and is unsuccessful. If the act is successful, then the first-order dispositions that led to the sender's signal and the receiver's predictive action are reinforced; otherwise, they may be weakened.

The agents' second-order dispositions determine how they learn by determining how they update the first-order dispositions that produce their signals and actions. Their second-order dispositions also determine what counts as success and failure in action. In particular, all it will mean to say that an action is successful is that it in fact leads to the reinforcement of whatever dispositions produced the action. Similarly, an action is unsuccessful if it does not lead to the reinforcement of the dispositions that produced the action.

Consider a senior vice president who attends a confidential board meeting for Corporation X every Tuesday at 10 am, observes whether the CEO nervously taps his pen on the mahogany boardroom table during the meeting, then leaves the meeting and can be seen walking to lunch at precisely 11:45 am either wearing his hat or not. His friend is a stockbroker. The broker observers whether the senior vice president is wearing his hat or not, then either buys stock in Corporation X for both himself and his friend or sells-short for both. See Figure 5.1.

Suppose the players' first-order dispositions to signal and act are determined as follows. The senior vice president has two urns. One urn is labeled "tap" and the other "no tap." Each urn starts with two balls, one labeled "hat" and the other "no hat." If the CEO taps his pen during the board meeting, the vice president draws a ball at random from the "tap" urn, then does what

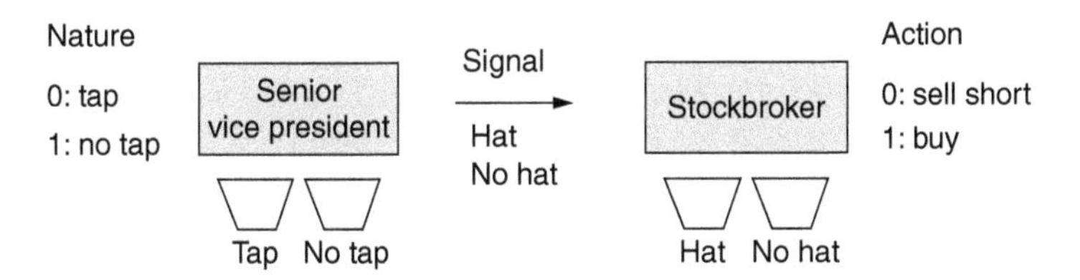

Figure 5.1 A one-sender, two-state game

it tells him to do regarding lunch attire; otherwise, he draws from the "no tap" urn and follows the instructions on that ball. The stockbroker also has two urns. One is labeled "hat" and the other is labeled "no hat." Each urn starts with two balls, one is labeled "buy" and the other "sell-short." If the broker sees his friend walking to lunch wearing his hat, he draws from the "hat" urn and does what it tells him to do; otherwise, he draws from the "no hat" urn.

The friends also have second-order dispositions that update how they signal and act from week to week. If the broker buys stock in Company X and the stock goes up or if he sells-short in Company X and the stock goes down, then each agent returns the ball he drew to the urn from which it was drawn and adds to that urn another ball of the same type; otherwise, each agent just returns the ball he drew to the urn from which it was drawn.[12]

Suppose, finally, that whether the value of Company X's stock goes up or down in a particular week is random and unbiased but that there is, in fact, a sure-fire natural correlation between the CEO tapping his pen and the change in value of the stock. In particular, the stock goes down when the CEO taps and goes up when he doesn't. Here, this is a law of nature that the agents must learn to be successful. But since the broker has no way of seeing what the CEO does during the meeting they will also have to coevolve a descriptive language.

This is an example of the simplest sort of sender–predictor game. As the game begins, the senior vice president wears his hat to lunch at random and the stockbroker buys and sells-short Company X stock at random. The vice president's hat-wearing behavior doesn't mean anything here, and the two friends, for what it's worth, can argue that no trades are being made on the basis of insider information. But as the senior vice president's dispositions to signal conditional on the state of nature and the broker's dispositions to act conditional on the signal evolve, the signals may become meaningful and the predictive actions may become successful.

Whether coordinated descriptive and predictive dispositions evolve in a particular game depends on the number and distribution of states of nature, the agents' signaling resources, the relationship between prior and posterior states of nature, and the precise second-order dispositions or learning

[12] This is an example of Herrnstein (1970) reinforcement learning, perhaps the simplest sort of learning one might consider. See Roth and Erev (1995), Erev and Roth (1998), and Barrett and Zollman (2009) for examples of more subtle types of reinforcement learning and how they relate to the game-playing behavior of human subjects. Reinforcement learning with punishment, considered below, is among these. See also Huttegger, Skyrms, Tarrès, and Wagner (2014) for a discussion of various learning dynamics in the context of signaling games with both common and conflicting interests.

dynamics one considers. One can prove that in the very simple game described here – a game with two states, two signals, two possible predictive actions, random unbiased nature, simple reinforcement learning, and a deterministic relationship between prior and posterior states – the senior vice president and broker are guaranteed to coevolve a perfectly descriptive language and perfectly successful coordinated predictive dispositions.[13]

While the senior vice president and the stockbroker may continue to maintain that they are just drawing balls from urns, acting, then putting the balls back, it is now significantly less plausible that they are innocent of trading on events that occur at the board meeting. Rather, at least on the face of it, they have coevolved a simple, but fully meaningful language and coordinated predictive dispositions which allow them to exploit a regularity in the CEO's behavior to their own advantage. The evidence of this is the uniform success of the broker's investments on the basis of the senior vice president's observations and signaling behavior.

In contrast with the Prisoner's Dilemma game with pre-play signaling, the agents here, in part because of their shared interests as represented by their coordinated second-order dispositions, end up with a stable descriptive language that is fully integrated with correspondingly stable predictive dispositions. Indeed, the relationship between description and prediction in this particular game is so tight that one might take the evolved hat-wearing behavior of the senior vice president to be either descriptive of what happened in the board meeting or prescriptive of what predictive action the broker should take. This easy symmetry in alternative interpretations of the evolved language, however, is not always exhibited by such games.

Suppose again that nature is deterministic but that there are four possible prior states of nature, each corresponding to a different posterior state, that are salient to the second-order dispositions of the agents. Suppose that in this game there are two senders, but that each sender has only two possible signals and hence each fails to have the resources required to fully specify the prior state of nature. Each sender has one urn for each possible prior state of nature, labeled 0, 1, 2, and 3. Each of their urns begins with one ball of each type of signal each sender might send, labeled 0 *and* 1. The predictor has one urn for each pair of signals the two senders might send labeled 00, 01, 10, and 11, where the first term corresponds to the signal from sender *A* and the second to

[13] Under these conditions, the sender–predictor game described here is equivalent to an unbiased $2 \times 2 \times 2$ Skyrms–Lewis signaling game. See Argiento et al. (2009) for the proof.

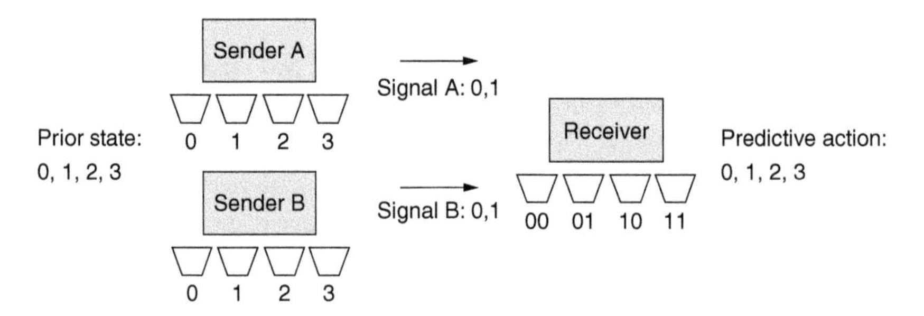

Figure 5.2 A two-sender, four-state game

that from sender *B*. Each of the predictor's urns begins with one ball of each type of predictive action he might take, labeled 0, 1, 2, and 3. See Figure 5.2.

In contrast with the simple reinforcement learning of the last game, we will suppose that the second-order learning dispositions of the inquirers are given by bounded reinforcement with punishment. More specifically, if the receiver's predictive action was successful, that is, if it matched the posterior state, then both of the senders and the predictor put the ball they drew back in the urn from which they drew it and add to that urn another ball of the same type if there are fewer than N_{max} balls of that type already in the urn. This will typically strengthen the dispositions that led to the predictive action. If the predictive action was unsuccessful, if it did not match the posterior state, then they each return the ball they drew to the urn from which they drew it if and only if it was the last ball of its type. This will typically weaken the dispositions that led to the predictive action. Note that, on these dynamics, there will always be at least one ball of each type in each urn, and never more than N_{max} balls of any particular type. Note also that there is no mechanism here for the senders to coordinate their signals directly. The only information they get regarding the other sender's behavior comes from their joint success and failure given the predictor's actions.

Here the senders and the predictor start by randomly signaling and randomly predicting, but, as they learn from experience, they typically evolve a set of nearly optimal, systematically intertwined, linguistic and predictive dispositions.[14] When they are successful, the senders evolve coordinated

[14] More specifically, on simulation, the senders and predictor are typically (0.995) found to evolve a set of nearly optimal (0.994) linguistic and predictive dispositions, and they spend most of their time near this state. But here there is no proof that such behavior is guaranteed. Indeed, here suboptimal partial pooling equilibria are often observed on simulation with the simple reinforcement learning described in the vice president–broker game (Barrett 2009).

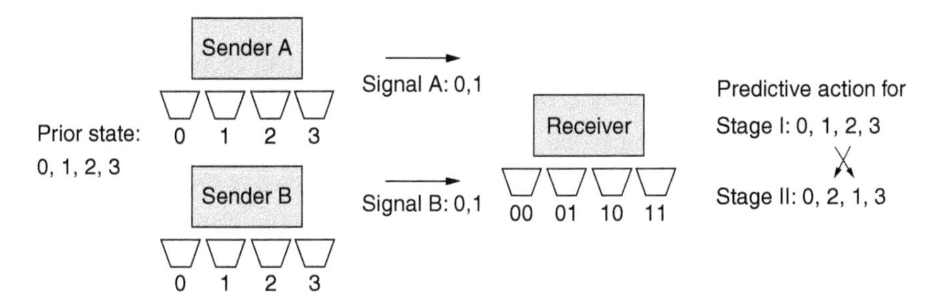

Figure 5.3 A two-sender, four-state game with changing reinforcement laws

partitions over the prior states of nature such that their signals might be thought of as providing information to the predictor regarding two distinct aspects of the prior state that jointly determine the posterior state.[15]

In contrast with the last game, it is less natural here to understand the senders as simply telling the receiver what predictive action to take since neither signal is sufficient to determine the receiver's action. While the agents' descriptive and predictive dispositions remain closely intertwined, considering how they respond to changes in nature further sharpens their roles in the game.[16]

Suppose that the way that the agents' first-order dispositions are updated changes after a successful set of descriptive and predictive dispositions have initially evolved. This might happen because the agents' norms, as represented by their second-order dispositions, change or it might happen because the relationship between prior and posterior states in nature changes. In either case, in the first stage of the game, the agents evolve successful descriptive and predictive dispositions just as described, then, in the second stage, they begin to reinforce posterior action 2 on prior state 1 and posterior action 1 on prior state 2 and weaken the dispositions associated with other actions involving these states. See Figure 5.3.

As before, the senders and predictor typically evolve successful descriptive and predictive dispositions in the first stage of the game. Then, in the second stage of the game, they typically adapt to the new reinforcements and evolve a

[15] The compositional language that evolves in this game is a very simple kind of language, in some ways similar to the compositional language used by putty-nosed monkeys or Campbell's monkeys. (See Arnold and Zuberbühler 2006, Arnold and Zuberbühler 2008, Ouattaraa et al. 2009.)

[16] See Barrett (2013a) and Barrett (2013b) for discussions of this and other games that might be taken to exhibit the coevolution of language and knowledge.

second set of nearly optimal descriptive and predictive dispositions.[17] But it is underdetermined precisely how the modeled inquirers will evolve the new set of dispositions. Sometimes the predictor evolves to make predictions differently, and sometimes, somewhat less frequently, the two senders evolve new coordinated partitions that classify the prior states of nature differently.[18]

When the senders evolve a new descriptive language, the result typically ends up looking something like a small-scale Kuhnian revolution, with the old and new descriptive languages being term-wise incommensurable.[19] On a particular run of the game, for example, the term 0_A might initially evolve to mean prior state 0 or 1 obtains, 1_A prior state 2 or 3, 0_B prior state 1 or 2, and 1_B prior state 0 or 3; then, in the second stage of the game, a new language might evolve where the term 0_A still means prior state 0 or 1 and 1_A still means prior state 2 or 3, but 0_B comes to mean prior state 1 or 3 and 1_B comes to mean prior state 0 or 2. Here, while both the old and new languages are equally descriptive faithful and lead to perfectly successful action, there is no term in the first language that translates 0_B in the second language. In this case, there is a term-wise incommensurability between the evolved languages, but they are statement-wise commensurable insofar as the senders can represent precisely the same states in each language; but it can also happen that sequentially evolved languages exhibit neither term-wise nor statement-wise commensurability.[20]

Successful inquiry here consists in the agents evolving toward a reflective equilibrium between their first-order dispositions to signal and predict, their second-order dispositions to reinforce their dispositions to signal and predict, and the results of their signals and predictions given the de facto nature of the

[17] More specifically, the agents are typically (0.972) found to evolve a second set of nearly optimal (0.993) descriptive and predictive dispositions that answer to the new regularities of nature or the agents' new second-order dispositions. The fact that they can unlearn their earlier dispositions on the dynamics is an important part of the story here.

[18] On simulation of the present game, the predictor evolves to make predictions differently about 0.58 of the time and the senders evolve new coordinated partitions about 0.42 of the time. It seems to be more evolutionarily difficult, then, for the senders to evolve a new coordinated language than for the receiver to evolve a new way of making predictions.

[19] See Kuhn's (1996) discussion of the incommensurability of descriptive languages in sequential theories. Of course, the present evolutionary model is vastly simpler than what one would find in real theory change.

[20] This might happen, for example, if the sequentially evolved languages correspond to different suboptimal partial pooling equilibria (Barrett 2009). While the agents in this case cannot individuate precisely the same states in both languages, their descriptions may be equally faithful in the overall precision to which they allow for the individuation of states and success in their predictive actions.

world. Their inquiry is ideally successful on the endogenous standard at hand when they no longer need to tune their descriptive language or how they use those descriptions to make predictions in order to satisfy their second-order dispositions.[21]

While it is the agents' second-order dispositions that drive inquiry by providing a dynamics for their first-order dispositions to signal and act, there is no reason to suppose that the agents' evolved dispositions somehow fail to represent objective knowledge concerning the world they inhabit. Indeed, successful inquiry here requires that they evolve dispositions that faithfully represent prior states of nature and reliably predict the posterior states determined by natural law. This is illustrated by the fact that if the agents are initially successful and one changes the regularities of the world they inhabit, they typically fail to be successful until they evolve a new set of first-order dispositions that track the new regularities and coordinate them with what the agents value as represented by their second-order dispositions.

5.4 Discussion

Just as sender–predictor games illustrate how agents may coevolve coordinated descriptive and predictive dispositions, the Prisoner's Dilemma with pre-play signaling shows how coordinated descriptive and strategic dispositions may coevolve in a population. Indeed, the agents' strategic dispositions in the pre-play signaling game are just predictive dispositions concerning the expected actions of opponents. Further, in both types of game, the agents' second-order dispositions provide endogenous standards for success and failure in their respective forms of inquiry.

Perhaps the most salient difference between the sender–predictor games and the Prisoner's Dilemma with pre-play signaling is that while the common interest of the agents in the former allows for convergence to an optimal set of descriptive and predictive dispositions, the conflicting interests of the agents in the latter promotes a continual arms race.

This instability in the evolving strategy types in the Prisoner's Dilemma with pre-play signaling is akin to the instability of the first-order dispositions of the agents in the sender–predictor game when one changes how the world behaves after the agents initially evolve successful descriptive and predictive

[21] There is a close relation between this endogenous notion of successful inquiry and a pragmatic notion of successful inquiry like that of C. S. Peirce. See, for example, chapters 3, 7, and 8 of Houser and Kloesel (1992).

dispositions. But in the Prisoner's Dilemma with pre-play signaling, it is the conflicting interests of the agents that leads to changes in precisely those features of the world that matter for the success of the agents' predictive actions.

As in the sender–predictor games, the descriptive languages that the agents evolve in the Prisoner's Dilemma with pre-play signaling are intimately intertwined with their coevolved predictive dispositions. Signals are descriptive of the strategic behaviors of the agents who use them, but here how they are descriptive evolves as signals are established between cooperators and then subverted by defectors.

In the sender–predictor games the object of knowledge is the stable regularities of the world the agents inhabit. In contrast, the object of knowledge in the Prisoner's Dilemma with pre-play signaling is the coevolving descriptive and predictive behavior of other agents.

Knowledge concerning the dispositions of competing agents here is knowledge concerning a moving target. Consequently, it can be expected to be both harder to come by and less stable than knowledge in service of common interests and concerning stable regularities.

The methodological moral here is that insofar as one believes that how we interact is the result of how our knowledge of each other and our strategies coevolve, evolutionary game theory provides the appropriate context in which to model social behavior generally and conditions for cooperation more specifically. Indeed, the sort of dynamical stories of interdependent dispositions that we have been telling can only be told in an evolutionary setting.

6 I cannot cheat on you after we talk

Cristina Bicchieri and Alessandro Sontuoso

6.1 Introduction

The experimental literature on social dilemmas has long documented the positive effect of communication on cooperation. Sally (1995), in a meta-analysis spanning thirty-five years of Prisoner's Dilemma experiments, shows that the possibility of communicating significantly increases cooperation. Social psychologists have explained such a finding by hypothesizing that the act of communicating contributes to promoting trust by creating empathy among participants (see Loomis (1959), Desforges et al. (1991), Davis and Perkowitz (1979)). Bicchieri (2002, 2006), in a different perspective, puts forward a *focusing function of communication* hypothesis, according to which communication can focus agents on shared rules of behavior and – when it does focus them on pro-social ones – generates a normative environment which is conducive to cooperation. More specifically, when individuals face an unfamiliar situation, they need cues to understand how best to act and, for this reason, they check whether some behavioral rule *they are aware of* applies to the specific interaction. The effect of communication is to make a behavioral rule *situationally salient*, that is, communication causes a shift in an individual's focus towards the strategies dictated by the now-salient rule. In doing so, communication also coordinates players' mutual expectations about which strategies will be chosen by the parties. In other words, (under some conditions) communication elicits social norms.

While a large proportion of studies on the effect of pre-play communication focuses on Prisoner's Dilemma games, Bicchieri, Lev-On, and Chavez (2010) examine behavior in sequential trust games. In what follows we shall look at those findings, discuss an interpretation based on the above hypothesis, and suggest a theoretical application that can account for it. Given that our analysis equally applies to the Prisoner's Dilemma, this essay contributes to the broad literature on social dilemmas by proposing an application for dynamic interactions.[1]

Bicchieri (2002) provides the basis for the focusing function of communication argument: when a rule of behavior becomes situationally salient, it causes a shift in an individual's focus, thereby generating empirical and normative expectations that direct one's actions. Before we elaborate on the argument, it is convenient to summarize Bicchieri's conditions for a social norm to exist and be followed.[2] A social norm exists and is followed by a population if two conditions are satisfied. First, every individual must be aware that she is in a situation in which a particular rule of behavior applies ("contingency" clause). Second, every individual prefers to conform to it, on the double condition ("conditional preference" clause) that: (1) she believes that most people conform to it (i.e. *empirical expectations* condition), and (2) she believes that most people believe she ought to conform to it (i.e. *normative expectations* condition).

In order to develop an equilibrium model that can capture more precisely the variables the experimenter manipulates in a laboratory environment, Bicchieri's model can be integrated with *psychological game theory* (Geanakoplos et al. 1989, Battigalli and Dufwenberg 2009). Such integration allows one to explicitly formalize the impact of the above conditions on a

[1] It should be noted that a dichotomous Trust Game can be thought of as a special version of the sequential Prisoner's Dilemma where payoffs are given as follows:

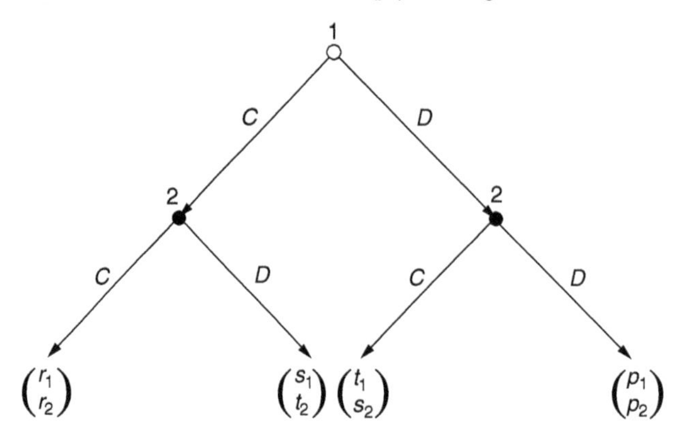

with $r_1 > t_1 = p_1 > s_1$ and $t_2 > r_2 > p_2 = s_2$. (Instead, in a Prisoner's Dilemma payoffs satisfy the following inequality: $t_i > r_i > p_i > s_i$, $\forall i \in \{1, 2\}$.)

[2] Bicchieri (2006: 11).

player's utility by incorporating conjectures about norms into the expected utility of a conformist player (Sontuoso 2013).

This chapter draws on the above theoretical treatments and provides an application illustrating how a formal framework that allows for different conjectures about norms is able to capture the focusing function of communication and to explain experimental results. In sum, the core of the argument is that *communication can focus people on some behavioral rule that is relevant to the specific interaction.* In so doing, it coordinates players' mutual expectations about which strategies will be played. So – if the aforementioned contingency condition holds – one may assume that making a behavioral rule salient through communication will result in greater compliance with the rule (Cialdini et al. 1991).

The remainder of the chapter is organized in this manner: Section 6.2 discusses Bicchieri, Lev-On, and Chavez's (2010) experimental results on the effect of communication in trust games; Section 6.3 briefly reviews models of norm compliance; Section 6.4 provides an application accounting for the focusing function of communication; and Section 6.5 draws conclusions.

6.2 Experimental evidence

Bicchieri, Lev-On, and Chavez (2010) study two features of pre-play communication in trust games: *content relevance* of the message (i.e. relevant vs. irrelevant communication) and *media richness* (i.e. face-to-face, "FtF," vs. computer-mediated communication, "CMC").

Consider the following trust game: the investor (first-mover) receives $6 and can choose to send any discrete amount x to the trustee (second-mover); the amount the trustee receives is tripled by the experimenter, so that the trustee can then send any discrete amount y in the interval $[0, 3x]$ back to the investor. See Figure 6.1.

In the experiment participants were paired randomly and played three variants of the above game, each time with a different subject, in the following order: (1) no-communication game (i.e. the baseline condition just described); (2) irrelevant or relevant CMC communication game; (3) irrelevant or relevant FtF communication game.[3] Investors did not receive feedback on the amount

[3] In the *CMC* conditions subjects could communicate via computer-based text chat for five minutes, whereas in the *FtF* conditions subjects communicated face-to-face for two minutes. In the *irrelevant* conditions subjects were instructed that they could discuss only the questions given by the experimenter (about a completely irrelevant topic), whereas in the *relevant*

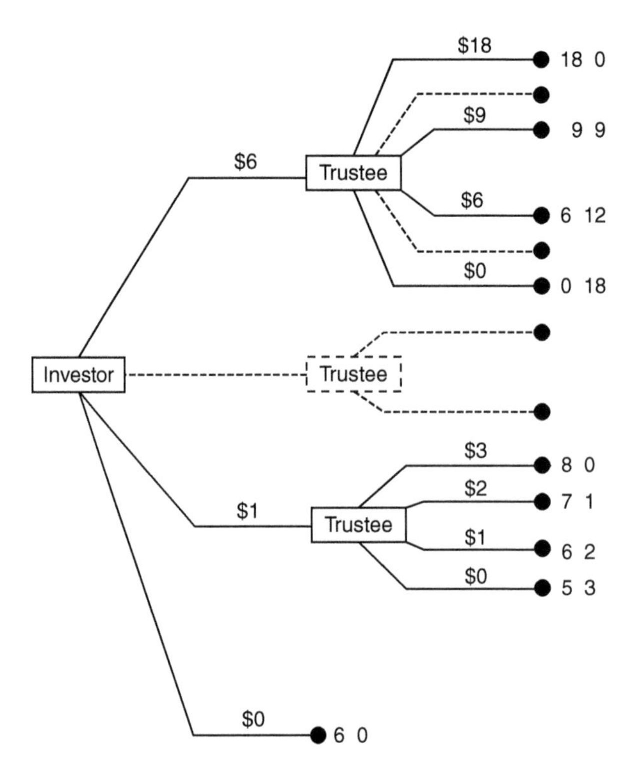

Figure 6.1 The trust game of Bicchieri, Lev-On, and Chavez (2010)

that the trustee returned until the end of the experimental session; also, after making their decision – in each variant of the game – investors were asked to report their expectation about the amount returned by the trustee. The authors were interested in three dependent variables: trust (defined as the amount of dollars sent by the investor), reciprocity (the amount returned by the trustee), and expectation (the amount the investor expected to get back).

Table 6.1 shows the participants' average responses across the five combinations of relevance and medium: a first look at the table reveals that both relevance and medium had large, positive effects on the three dependent variables. Note that *relevant face-to-face* communication had the largest effects on all three variables while *relevant computer-mediated* communication had the second largest effects. Figure 6.2 on p.106 illustrates the distribution of trust across the five conditions.

conditions they could discuss any topic except their identities. (Roughly half of the experimental sessions featured the relevant conditions while the remaining sessions featured the irrelevant conditions.)

Table 6.1 Mean (SEM) of trust, reciprocity, and investor's expectations by communication relevance and medium ($N = 96$). FtF = Face-to-face; CMC = Computer-mediated communication.

	Control (N=32)	FtF-Relevant (N=14)	CMC-Relevant (N=14)	FtF-Irrelevant (N=18)	CMC-Irrelevant (N=18)
Trust	2.63 (0.36)	5.57 (0.46)	5.14 (0.57)	4.17 (0.49)	3.28 (0.61)
Reciprocity	1.92 (0.48)	7.57 (0.96)	5.14 (1.33)	3.33 (1.05)	1.94 (0.78)
Expected reciprocity	3.54 (0.53)	8.36 (0.69)	7.43 (0.96)	5.56 (0.91)	4.28 (0.93)

As shown in Figure 6.2, investors were most trusting in both *relevant* communication conditions (where the majority sent their entire $6 endowment).

The authors' further data analysis discloses the following key points: (1) the behavior of investors was strongly determined by their *expectations of trustees' reciprocation*;[4] (2) the variable most conducive to creating such expectations was not the medium, but rather the *content relevance* of the message (i.e. investments were significantly higher following unrestricted communication than restricted or no communication and – whenever communication was restricted to irrelevant topics – there were no significant differences between the amounts sent in the CMC, FtF, and control conditions); (3) reciprocity significantly increased with *trust*, *content relevance* of the message, and *medium* (more precisely, reciprocity was higher in the CMC condition for lower amounts of trust but became higher in the FtF condition for higher amounts of trust).[5]

How do such results relate to the two explanations for the effect of communication on social dilemmas that we mentioned in the introduction (i.e. one explanation maintains that communication enhances cohesion and

[4] Note, however, that those expectations were rarely met, since expected reciprocation was significantly higher than the actual reciprocation across conditions (except when $6 was invested).

[5] For instance, as the amount that the investor sent approached zero, the odds that the trustee returned each available dollar were over seven times higher in CMC than in FtF. With each additional dollar that the investor sent, however, the odds that the trustee reciprocated increased more rapidly in FtF conditions. In other words, the probability of returning each available dollar increased with the amount invested, but increased more rapidly for the FtF and control conditions than for CMC.

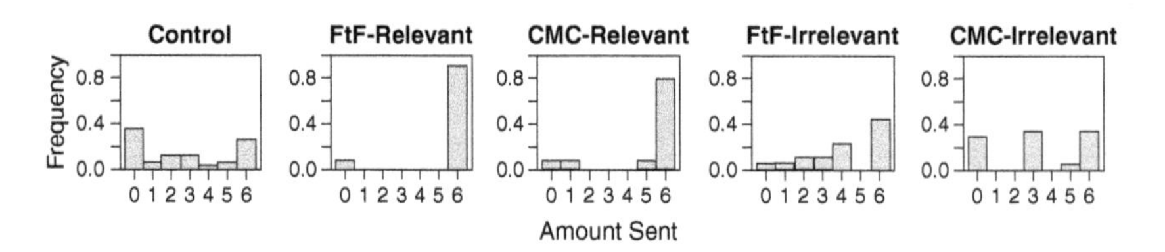

Figure 6.2 Distribution of trust by communication medium and relevance

group identity while the other asserts that communication elicits social norms)? The data seem to provide evidence in favor of the latter explanation: in fact, if the former were valid, then one should not find an effect of content relevance on the expected reciprocation (point (2) above). On the other hand, Bicchieri's focus theory of norms is consistent with the data, since it predicts an effect of the message content relevance on the expected reciprocation. Specifically, Bicchieri (2002, 2006) hypothesizes that, when participants are allowed to talk about the strategic situation at hand, the discussion on "how to appropriately behave" will lead the participants to become aware of the fact that the current interaction is one to which some default rule of behavior applies.[6] Hence, focusing subjects on a rule of behavior generates and coordinates empirical and normative expectations.

6.3 Theoretical foundations

In what follows we review two formal, theoretical treatments of norms and subsequently draw on them to develop an application accounting for the experimental results.

Bicchieri (2006: 52) proposes a general utility function based on norms. Considering an n-player normal form game, let S_i denote the strategy set of Player i and $S_{-i} = \prod_{j \neq i} S_j$ be the set of strategy profiles of players other than i. A norm N_i is defined as a (set-valued) function from one's expectation about the opponents' strategies to the "strategies one ought to take," that is, N_i: $L_{-i} \rightarrow S_i$, with $L_{-i} \subseteq S_{-i}$.[7] A strategy profile $s = (s_1, \ldots, s_n)$ is said to *instantiate* a norm for Player j if $s_{-j} \in L_{-j}$ (i.e. if N_j is defined at s_{-j}), and

[6] See Lev-On et al. (2010) for the effect of group (vs. dyadic) communication in trust games.

[7] For example, in an n-player Prisoner's Dilemma a shared norm may be to cooperate: in that case, L_{-i} includes the *cooperate* strategies of all players other than i. Note that in the case where – given the others' strategies – there is not a norm prescribing how Player i should behave, then N_i is not defined at L_{-i}.

to *violate* a norm if, for some j, it instantiates a norm for j but $s_j \neq N_j(s_{-j})$. Player i's utility function is a linear combination of i's material payoff $\pi_i(s)$ and a component that depends on norm compliance:

$$(1) \quad U_i(s) = \pi_i(s) - k_i \max_{s_{-j} \in L_{-j}} \max_{m \neq j} \{\pi_m(s_{-j}, N_j(s_{-j})) - \pi_m(s), 0\}$$

where $k_i \geq 0$ represents i's sensitivity to the norm and j refers to the norm violator. The norm-based component represents the maximum loss (suffered by players other than the norm violator j) resulting from all norm violations: the first maximum operator takes care of the possibility that there might be multiple rule-complying strategy profiles; the second maximum operator ranges over all the players other than the norm violator j. Bicchieri's utility function makes it possible for the experimenter to test whether subjects' behavior is consistent with preferences for conformity to a social norm, given that the above-mentioned conditions for the existence of a social norm are satisfied (i.e. contingency, and preferences conditional on the relevant empirical and normative expectations; see Bicchieri 2006). Specifically, this utility function captures conformist preferences in case a norm exists. Hence, the norm-based component of the utility function represents the maximum loss resulting from all violations of an *established* norm.

Bicchieri's utility function makes very sharp predictions in cases where there is no ambiguity about subjects' expectations as to what the norm prescribes. In order to explicitly represent conditionally conformist preferences in dynamic games where multiple rules of behavior may apply, Sontuoso (2013) extended Bicchieri's framework to a "psychological" utility function and a belief-based formulation of her conditions for norm existence (which are directly reflected into the player's utility function). Given that in Bicchieri's theory of norms *expectations are crucial to compliance*, having a model of how subjects derive their conjectures about norms may be useful for interpreting the experimental results of Bicchieri, Lev-On, and Chavez (2010).

Before outlining such a model of norm compliance, we shall introduce some notation on dynamic games: let an extensive form game be given by $\langle N, H, P, (I_i)_{i \in N} \rangle$, where N is the set of players, H is the set of feasible histories, I_i is the information partition of Player i. Further, let Z denote the set of terminal histories, with $H \setminus Z$ being the set of non-terminal histories; given that, let P denote the player function (which assigns to each element of $H \setminus Z$ an element of N), and let $A_i(h)$ denote the set of feasible actions for

Player i at history h.[8] The *material payoffs* of players' strategies are described by functions m_i: $Z \to \mathbb{R}$ for each player $i \in N$. Then, denote the set of Player i's pure strategies allowing history h as $S_i(h)$; strategy profiles allowing history h are defined as $S(h)$, and $S_{-i}(h)$ for all players j other than i; given that, let $z(s)$ indicate a terminal history induced by some strategy profile $s \in S$. Battigalli and Dufwenberg (2009) provide a framework for the analysis of *dynamic psychological games* where conditional higher-order systems of beliefs influence players' preferences: as in their model, here it is assumed that (at each history) every player holds an updated system of first-order beliefs $\alpha_i = (\alpha_i(\cdot|h))_{h \in H_i}$ about the strategies of all the co-players;[9] at each history Player i further holds a system of second-order beliefs β_i about the first-order belief system of each of the opponents.[10]

The model of norm compliance we employ (Sontuoso 2013) assumes there exists a set of default rules of behavior, where each rule specifies strategies appropriate to generic (mixed-motive) games; players have a subset of such rules stored in their minds and derive from them "norm-conjectures" (i.e. expectations as to which rule-driven strategies apply to the current game). Therefore, if an individual j is a norm-driven player – and presumes that her co-players are norm-driven too – she can form her first-order belief α_j by assuming her co-players' behavior to be consistent with some rule. A *"behavioral rule"* is defined as a set-valued function r that assigns to every non-terminal history h one or more elements from the set of strategy profiles $S(h)$.[11] The *set of behavioral rules* is denoted by R, and the *behavioral rule subset of Player i* by R_i (with $R_i \subseteq R$), with R_i representing the set of rules i is aware of. Further, given a game G and some rule $\dot{r}$, one can derive the

[8] Note that a node of the game tree is identified with the history leading up to it (i.e. a path in the game tree) as in Osborne and Rubinstein (1994).

[9] For example, in a game with perfect information, at each $h \in H_i$ Player i holds an updated belief $\alpha_i(\cdot|h)$ such that she believes that all players have chosen all the actions leading to h with probability 1.

[10] It is assumed that players' beliefs at different information sets must satisfy Bayes' rule and common knowledge of Bayesian updating.

[11] For instance, consider a rule that prescribes behavior minimizing payoff-inequality among players: when one evaluates this rule at the initial history, the rule will dictate those strategy profiles that minimize the difference in payoffs among players, considering that every terminal node can be reached; instead if one of the players deviates along the play, when evaluating this behavioral rule at a history following such a deviation, the rule will dictate strategy profiles that minimize the difference in payoffs among players, conditional on the terminal nodes that can still be reached (Sontuoso 2013).

set of strategy profiles dictated by $\dot{r}$ (e.g. the set of strategy profiles dictated by $\dot{r}$, when evaluated at the initial history, is denoted by $\dot{r}(h^0)$); given $R_i \subseteq R$, one can derive the *set of Player i's "rule-complying" actions at history h* (denoted by $A_{i,h}(R_i(h^0))$), which depicts the set of actions prescribed – by any of the rules $r \in R_i$ – to Player i at history h.

Finally, a *"norm-conjecture"* of Player i is defined as a collection of independent probability measures $\rho_i = (\rho_i(\cdot|h))_{h\in H\setminus Z}$ (with $\rho_i(a|h)$ being the probability of action a at history h) such that *the support of* ρ_i *is a subset of the rule-complying actions of the active player at history h.* Conditionally conformist preferences are represented by an expected utility function given by a linear combination of the player's material payoff and a component representing some disutility arising from deviations from the presumed norm.[12] Formally, a norm-driven individual has conditionally conformist preferences characterized by a utility function u_i^C of the form

$$(2)\quad u_i^C(z,s_{-i},\alpha_j) = m_i(z) - k_i d_i^C d_i^E\left(1 + \sum\nolimits_{j\neq i} max\{0, E_{\rho_i,s_j,\alpha_j}[m_j|h^0] - m_j(z)\}\right),^{13}$$

with $s_{-i} \in S_{-i}(z)$, $k_i \in [0, \infty)$ and where: k_i is Player i's *sensitivity to the presumed norm*; d_i^C is a dummy variable equal to one if i is aware of one or more behavioral rules applicable to the given game, equal to zero otherwise; d_i^E is a dummy variable equal to one if i believes that every $j \neq i$ is aware and will also adhere to some $r \in R$, equal to zero otherwise.[14]

In the next section we shall provide an application illustrating how a formal framework that can allow for different (conjectures about) norms is able to capture the focusing function of communication and to explain the experimental results of Bicchieri, Lev-On, and Chavez (2010).

[12] More precisely, the anticipated disutility is a function of any positive difference between the "initially expected payoff to j" and the payoff j would get in the event of a rule violation. Note that it is assumed that if Player j is a norm-driven individual – and presumes that her co-players are norm-driven too – she can form her first-order belief α_j by assuming her co-players' behavior to be consistent with some norm-conjecture $\rho_j = \rho_i$ (hence, with some rule r).

[13] $E[X]$ denotes the expected value of X.

[14] After deriving a belief-based formulation of Bicchieri's conditions for a social norm to exist and be followed, Sontuoso (2013) proposed a notion of *"Social Sequential Equilibrium"* allowing for belief-dependent conformist motivations (by refining Battigalli and Dufwenberg's specification of the sequential equilibrium concept of Kreps and Wilson 1982).

6.4 An application accounting for the focusing function of communication

First, we shall define some specific behavioral rules reflecting principles which are usually assumed to regulate behavior in social dilemmas and which, one may assume, could apply to the trust game of Bicchieri, Lev-On, and Chavez (see Figure 6.1 above). It is useful to recall here the above definition of *behavioral rule*, i.e. a correspondence dictating the strategy profiles most "appropriate" – according to a certain principle – for each node of the given mixed-motive game. Also, given a set of potential rules R, we assume that each player's *culture identifies* a subset R_i (stored in i's memory) which contains default rules of behavior that the player is aware of (Sontuoso 2013).

Recall that the trust game of Bicchieri, Lev-On, and Chavez (2010) was defined as follows: the investor receives \$6 and can choose to send any discrete amount x to the trustee; the amount the trustee receives is tripled by the experimenter, so that the trustee can then send any discrete amount y in the interval $[0, 3x]$ back to the investor. Given that, here are some rules applicable to the experimental trust game.

- "Inequality-Reducing" rule, r^F: *any strategy profile such that the investor chooses an amount x (other than \$0), and the trustee chooses an amount y that minimizes the difference in payoffs.*
- "Pareto-Efficiency" rule, r^P: *any strategy profile such that the investor chooses \$6, and the trustee chooses any amount.*
- "Reciprocity" rule, r^C: *any strategy profile such that the investor chooses an amount x (other than \$0), and the trustee chooses an amount* $y \geq x$.

Before applying the above rules, note that in what follows we denote an *action* by the amount sent; for example, if the investor (I) chooses action \$6, then the set of the trustee's (T) feasible actions at history h = \$6 is given by $A_T(\$6) = \{\$18, \$17, \$16, \ldots, \$2, \$1, \$0\}$. It follows that if the investor chooses, say, \$1 and then the trustee chooses also \$1, the payoff profile induced by the path (\$1, \$1·) is given by the pair (\$6, \$2).

Now, considering for example r^F, one can derive the *set of strategy profiles dictated by* r^F which – when evaluated at the initial history – contains five elements, that is: (1) the strategy profile where the investor chooses \$6 and the trustee chooses $\$9_{if\ \$6}\ \$7_{if\ \$5}\ \$5_{if\ \$4}\ \$3_{if\ \$3}\ \$1_{if\ \$2}\ \$0_{if\ \$1}$, which yields the payoff profile (\$9, \$9); (2) the strategy profile where the investor chooses \$5 and the trustee chooses the same as above, which yields the payoff profile (\$8, \$8), etc. ... In short, using the above notation, the set of paths dictated by r^F is given by:

$$r^F(h^0) = \{(\$6, \$9\cdot), (\$5, \$7\cdot), (\$4, \$5\cdot), (\$3, \$3\cdot), (\$2, \$1\cdot)\}.$$

Similarly, considering r^P, one can derive the *set of paths dictated by* r^P which – when evaluated at the initial history – contains nineteen elements, that is:

$$\{(\$6, \$18\cdot), (\$6, \$17\cdot), \ldots, (\$6, \$1\cdot), (\$6, \$0\cdot)\}.$$

Also, the *set of paths dictated by* r^C is given by:

$$\left\{ \begin{array}{c} (\$6, \$18\cdot), (\$6, \$17\cdot), \ldots, (\$6, \$6\cdot), (\$5, \$15\cdot), (\$5, \$14\cdot), \ldots, (\$5, \$5\cdot), \\ (\$4, \$12\cdot), (\$4, \$11\cdot), \ldots, (\$4, \$4\cdot), (\$3, \$9\cdot), (\$3, \$8\cdot), \ldots, (\$3, \$3\cdot), \\ (\$2, \$6\cdot), (\$2, \$5\cdot), \ldots, (\$2, \$2\cdot), (\$1, \$3\cdot), (\$1, \$2\cdot), (\$1, \$1\cdot) \end{array} \right\}.$$

Next, if one assumes that both the investor (I) and the trustee (T) are aware of all the above behavioral rules (i.e. $R_I = R_T = \{r^F, r^P, r^C\}$) then one can derive, for each player, the *set of rule-complying actions at history h*, which depicts the set of actions prescribed – by any of the rules $r \in R_i$ – to Player i at history h. For the investor this is given by $A_{i=I,\ h=h^0}(R_i(h^0)) = \{\$6, \$5, \ldots, \$1\}$ while for the trustee there will be one set of rule-complying actions for each history, i.e.

$$A_{i=T,\ h=\$6}(R_i(h^0)) = \{\$18, \$17, \ldots, \$6\},\ A_{i=T,\ h=\$5}(R_i(h^0)) = \{\$15, \$14, \ldots, \$5\}, \ldots, A_{i=T,\ h=\$1}(R_i(h^0)) = \{\$3, \$2, \$1\}.$$

It is now clear that the aforementioned rules, when applied to the experimental trust game, dictate several strategy profiles. It then follows that the support of i's *norm-conjecture* ρ_i (i.e. the set of the active player's rule-complying actions that are assigned positive probability by ρ_i) may contain any of the above rule-complying actions. Especially in cases like this, where there are several admissible (i.e. rule-complying) actions – unless players can communicate – it might be difficult for them to engage in a process of mutually consistent belief formation relative to a presumed social norm: this may result in no social norm being followed. Instead, assume that a rule prescribing behavior that minimizes payoff-inequality among players is made salient *through communication*; in this case R_I and R_T are still defined as above (i.e. $R_I = R_T = \{r^F, r^P, r^C\}$), but now it is reasonable to assume that players will converge towards a norm-conjecture derived from r^F only (and, in turn, they will derive their first- and second-order beliefs from such a norm-conjecture). In light of the experimental results discussed in Section 6.2, it seems reasonable to conclude that – in the experiment of Bicchieri, Lev-On, and Chavez (2010) and, specifically, in the *relevant face-to-face* communication condition – the Inequality-Reducing rule r^F constituted a social norm and was being followed by the experimental subjects.[15] In fact, as shown in Figure 6.2, the relevant FtF communication game exhibited such a high level of trust that almost every investor contributed her entire \$6 endowment (while the average amount returned by the trustee was almost \$8, and the modal choice was \$9).

More explicitly, note that the norm-conjecture induced by r^F, for $\forall i \in N$, is such that: ρ_i may take on value 1 for any one of the investor's actions (other than \$0), and takes on value 0 for all of the trustee's actions but $\$9_{if\ \$6}$, $\$7_{if\ \$5}$, $\$5_{if\ \$4}$, $\$3_{if\ \$3}$, $\$1_{if\ \$2}$, $\$0_{if\ \$1}$. Given that, each player i can form her first-order belief α_i by assuming her co-player's behavior to be consistent with her norm-conjecture ρ_i; for example, the investor's belief $\alpha_I = (\cdot|h^0)$ will correspond to a probability measure over the strategies of the trustee, with the support of α_I containing only the opponent's rule-complying strategies. Using formula (2) above, the investor can then calculate the expected utility from each of her actions as follows: first, note that the investor's utility would involve a potential loss (i.e. a "psychological" disutility) only at x = \$0 while it would be maximized at x = \$6 (conditional on the trustee's sensitivity k_T). Yet before considering the latter case, let's look at the game from the trustee's perspective: in order to calculate the optimal action at each history after which she has to move, the trustee will compare her utility of conforming with her utility of deviating from the presumed norm; so, the *trustee's expected utility from deviating* (choosing, say, \$0) *after the investor has chosen* x = \$6 would equal $u_T^C(z, \rho_i, \beta_T) = 18 - k_T[1 + 9]$;[16] instead, the trustee's utility from choosing

[15] A social norm r^* (exists and) *is followed* by population N if: every player $i \in N$ has conformist preferences represented by a utility function u_i^C (as given in formula (2) above), with $d_i^C = 1$, $d_i^E = 1$, and $k_i > 0$; every player i maximizes her expectation of u_i^C; every i holds correct beliefs about every j's ($j \in N$, with $j \neq i$) first-order belief and behavior; every player i's behavior is consistent with one of the end-nodes yielded by $r^* \in R_i \bigcap R_j$ (according to norm-conjectures $\rho_j = \rho_i$ for $\forall j \in N$); k_i is sufficiently large for every $i \in N$ (Sontuoso 2013). (See also Bicchieri (2006: 11) for the conceptual distinction between the existence of a norm *versus* its being followed.)

[16] Note that, for some player i, $u_i^C(z, s_{-i}, \beta_i)$ represents i's estimation of $u_i^C(z, s_{-i}, \alpha_j)$.

\$9 after x = \$6 would simply correspond to her material payoff (i.e. $m_T(z(\$6, \$9)) = \$9$). In brief, the trustee's conformist preferences can be expressed compactly as: $18 - 10k_T \leq 9 \Rightarrow k_T \geq \frac{9}{10}$. If the investor effectively believes that $k_T \geq \frac{9}{10}$, then she will compare her utility from taking a rule-complying strategy against her utility from deviating from the presumed norm: the investor's expected utility from deviating (i.e. choosing x = \$0) would equal $u_I^C(z, \rho_i, \beta_I) = 6 - k_I[1 + \bar{y}]$, where the "initially expected payoff to the trustee" given the strategy $s_T = y\cdot$ (i.e. $E_{\rho_i, s_T, \alpha_T}[m_T|h^0]$) is now denoted by $\bar{y}$ for convenience. Instead, the investor's expected utility from choosing x = \$6 would be given by $m_I(z(\$6, \$9)) = \$9$. Hence, the investor's conformist preferences can be expressed compactly as: $6 - k_I[1 + \bar{y}] \leq 9 \Rightarrow k_I \geq -\frac{3}{1+\bar{y}}$, which is always satisfied. To conclude, the investor will choose \$6 and the trustee will choose $\$9_{if\ \$6}$ $\$7_{if\ \$5}$ $\$5_{if\ \$4}$ $\$3_{if\ \$3}$ $\$1_{if\ \$2}$ $\$0_{if\ \$1}$ (whenever $k_T \geq \frac{9}{10}$).

To sum up, if a rule prescribing behavior that minimizes payoff-inequality among players is made salient through communication, the equilibrium path will involve the investor choosing \$6 and the trustee choosing \$9, so that both players' payoff is \$9: again, looking at Table 6.1 above, one may conclude that – in the relevant face-to-face communication condition – subjects did play an equilibrium where norm-conjectures were correct. In other words, communication coordinated players' expectations by making a particular rule salient. Besides, note that moving from *relevant face-to-face* communication to *relevant computer-mediated* communication somewhat reduced both reciprocity and expected reciprocity. Indeed, when relevant communication is allowed, the expectation that trustees will abide by the social norm activated through communication is likely to be *less vivid in a computer-mediated* than in a face-to-face environment (and coordinating expectations will be more difficult).[17]

6.5 Conclusion

We investigated the focusing (and coordinating) function of communication: drawing on experimental evidence we have argued that – when a behavioral rule becomes situationally salient – it causes a shift in an individual's focus, thereby generating (empirical and normative) expectations that direct one's strategies. We presented an application illustrating how a formal framework

[17] In this respect see also the discussion in Bicchieri and Lev-On (2007).

that allows for different conjectures about norms is able to capture such a focusing function.

Given that this framework allows us to compare predictions under *different focal rules of behavior*, it could be of help in designing novel experiments that study the dynamics of the cognitive processes characterizing sequential trust games and, more generally, social dilemmas that involve a stage of non-binding, pre-play communication among participants.

7 Prisoner's Dilemma cannot be a Newcomb Problem

José Luis Bermúdez

Many philosophers, decision theorists, and game theorists have commented on apparent structural parallels between the Prisoner's Dilemma and Newcomb's Problem. David Lewis famously argued that to all intents and purposes there is really just one dilemma in two different forms: "Considered as puzzles about rationality, or disagreements between two conceptions thereof, they are one and the same problem. Prisoner's Dilemma is a Newcomb Problem – or rather, two Newcomb Problems side by side, one per prisoner. Only the inessential trappings are different" (Lewis 1979: 235).

This chapter explores and rejects Lewis's identity claim. The first section reviews the Prisoner's Dilemma and Newcomb's Problem. Section 7.2 explores the theoretical background to Lewis's argument, focusing in particular on how it has been taken to support causal decision theory over standard (evidential) decision theory. Section 7.3 sets out Lewis's argument. Section 7.4 identifies the flaw in Lewis's argument, while Section 7.5 compares the argument in 7.4 to the so-called symmetry argument in favor of cooperating in the Prisoner's Dilemma. Section 7.6 generalizes the discussion by showing that the two problems involve fundamentally different types of reasoning (parametric in one case, and strategic in the other), and hence that the Prisoner's Dilemma cannot be a Newcomb Problem.

7.1 Two puzzles about rationality

Figure 7.1 shows the payoff table in a standard version of the Prisoner's Dilemma. The payoffs are years in jail. Assuming that A and B both prefer to spend the least possible amount of time in jail, we can represent the payoff table from each player's perspective as follows, where A's preference ordering is given by $a_1 > a_2 > a_3 > a_4$ and B's preference ordering is given by $b_1 > b_2 > b_3 > b_4$. See Figure 7.2.

	B cooperates	B does not cooperate
A cooperates	1 year, 1 year	10 years, 0 years
A does not cooperate	0 years, 10 years	6 years, 6 years

Figure 7.1 The Prisoner's Dilemma

	B cooperates	B does not cooperate
A cooperates	a_2 , b_2	a_4 , b_1
A does not cooperate	a_1 , b_4	a_3 , b_3

Figure 7.2 The Prisoner's Dilemma: general version

This gives the classic structure of a Prisoner's Dilemma. Both players agree that mutual cooperation is the second-best outcome and mutual non-cooperation the third best outcome. But A's best outcome is B's worst, and B's best is A's worst.

If we assume that Prisoner's Dilemma is a one-off interaction, then the standard analysis is that it is rational not to cooperate. The reasoning is simply that non-cooperation dominates cooperation. Suppose that you are player A. Then there are two possible circumstances. The first is that B cooperates. But if B cooperates, then non-cooperation (a_1) is strictly preferred to cooperation (a_2). In the second circumstance, B does not cooperate. But here too non-cooperation (a_3) is strictly preferred to cooperation (a_4). So the rational thing for player A to do is not to cooperate. Things are exactly similar for player B, and so a rational player B will not cooperate either. With both players choosing rationally, therefore, the outcome is mutual non-cooperation. This is a strong Nash equilibrium because neither player can improve their position by unilaterally changing their choice.

Turning to Newcomb's Problem (henceforth: NP): As it is standardly presented, you are faced with a Predictor whom you have good reason to believe to be highly reliable. The Predictor has placed two boxes in front of you – one opaque and one transparent. You have to choose between taking just the opaque box (*one-boxing*), or taking both boxes (*two-boxing*). You can

	The Predictor has predicted two-boxing and so the opaque box is empty	The Predictor has predicted one-boxing and so the opaque box contains $1,000,000
Take just the opaque box	$0	$1,000,000
Take both boxes	$1,000	$1,001,000

Figure 7.3 Newcomb's Problem

see that the transparent box contains $1,000. The Predictor informs you that the opaque box may contain $1,000,000 or it may be empty, depending on how she has predicted you will choose. The opaque box contains $1,000,000 if the Predictor has predicted that you will take only the opaque box. But if the Predictor has predicted that you will take both boxes, the opaque box is empty.

The pay-off table in Figure 7.3 shows that dominance reasoning applies here also. There are two circumstances. Either there is $1,000,000 in the opaque box or there is not. In either case you are better off two-boxing than one-boxing.

NP has been much discussed by decision theorists because this dominance reasoning seems to conflict with the requirements of maximizing expected utility. Suppose you assign a degree of belief of 0.99 to the Predictor's reliability (and suppose that expected utility in this case coincides with expected monetary value). Then the expected utility of one-boxing is 0.99 x $1,000,000 = $990,000, while the expected utility of two-boxing is (0.99 x $1,000) + (0.01 x $1,001,000) = $11,000. So, a rational agent following the MEU principle (the principle of maximizing expected utility) will take just the opaque box.

Many philosophers and some decision theorists have argued that the apparent conflict in NP between dominance and the principle of maximizing expected utility shows that we need to rethink the principle.[1] What NP shows, they claim, is the inadequacy of thinking about expected utility in purely

[1] See, e.g., Nozick (1969), Gibbard and Harper (1978), Skyrms (1980), Lewis (1981). For dissenting views, see Horgan 1981 and Eells (1981).

evidential terms. On the evidential approach, the expected utility of a given action is a function of the conditional probabilities of the different outcomes of that action. Instead, these theorists propose different versions of causal decision theory, where calculations of expected utility are based on probability calculations that track causal relations between actions and outcomes, as opposed simply to probabilistic dependence relations between actions and outcomes.

The basic idea is that what matters when one is thinking about how it is rational to choose are the effects that one's choice will bring about. It is true that there is a high probability of there being nothing in the opaque box if I two-box, but my two-boxing is causally independent of the prediction that leads to there being nothing in the opaque box, and so causal decision theorists think that it should not be relevant. From the point of view of causal decision theory, when an outcome O is causally independent of an action a, then $\Pr(O/a)$, the probability of O conditional upon a, is the same as $\Pr(O)$, the unconditional probability of O.

The application to NP is as follows. Let O_1 represent the outcome of the Predictor predicting two-boxing and so putting nothing in the opaque box, and let O_2 represent the outcome of the Predictor predicting one-boxing and so putting \$1,000,000 in the opaque box. Let $\Pr_C(O_1/a)$ represent the causal probability of O_1 conditional upon action a. Then we have

$$\Pr_C(O_1/\textit{one-boxing}) = \Pr_C(O_1/\textit{two-boxing}) = \Pr(O_1) = x$$

and

$$\Pr_C(O_2/\textit{one-boxing}) = \Pr_C(O_2/\textit{two-boxing}) = \Pr(O_2) = y.$$

The causal expected utility (CEU) calculations therefore come out as:

$$\text{CEU}(\textit{one-boxing}) = 0x + \$1{,}000{,}000y = \$1{,}000{,}000y$$
$$\text{CEU}(\textit{two-boxing}) = \$1{,}000x + \$1{,}000{,}000y$$

Plainly, the principle of maximizing causal expected utility prescribes two-boxing and so is consistent with the dominance reasoning.

7.2 Comparing the two puzzles

Even from this brief survey it is clear that there are two prima facie similarities between NP and the Prisoner's Dilemma. In particular,

(1) In both NP and the Prisoner's Dilemma outcomes are causally independent of actions.
(2) In both NP and the Prisoner's Dilemma there is a dominant action (two-boxing and non-cooperation respectively).

The principal focus of this chapter is David Lewis's well-known and widely accepted argument that there is much more than prima facie similarity between NP and the Prisoner's Dilemma (Lewis 1979). According to Lewis, NP and the Prisoner's Dilemma are really notational variants of each other – two different ways of packaging a single problem. We will be looking in detail at Lewis's argument in later sections. In this section I will discuss the significance of Lewis's argument and some of the principal reactions in the literature.

As was noted in the previous section, NP has been taken to support causal decision theory. It is natural to respond, however, that NP is too far-fetched to provide an effective challenge to standard, evidential decision theory. Hard cases make bad law, and decision problems involving nearly infallible predictors are a poor basis for overturning a theory that is mathematically well-grounded, foundational to the social sciences, and arguably captures the core of intuitive notions of rationality.[2] NP is interesting because it shows that in highly artificial cases we can tease apart probabilistic dependence and causal dependence by constructing scenarios where outcomes are probabilistically dependent upon actions, but causally independent of those actions. But at the same time it is too distant from ordinary decision-making to require any revision to standard and well-established ways of thinking about expected utility.

If Lewis's argument is sound, however, this objection misses the mark completely. It is widely accepted that Prisoner's Dilemma-type interactions are ubiquitous in human interactions (see Axelrod 1984 for interesting examples and discussion of "real-life" Prisoner's Dilemmas). So, if the Prisoner's Dilemma is simply a notational variant of NP, then we potentially have a much stronger base for attacking evidential decision theory. And that is precisely Lewis's point. Perhaps the most significant conclusion he draws from his discussion of NP and the Prisoner's Dilemma is that there is really no difference between two-boxing in NP and non-cooperation in a one-shot Prisoner's Dilemma. The two problems are really equivalent and so too are

[2] For more discussion of the relation between decision theory and rationality, see Bermúdez (2009).

these two strategies. Here is how Lewis puts it in the closing two paragraphs of his paper:

> Some – I, for one – who discuss Newcomb's Problem think it is rational to take the thousand no matter how reliable the predictive process may be. Our reason is that one thereby gets a thousand more than he would if he declined, since he would get his million or not regardless of whether he took his thousand. And some – I, for one – who discuss Prisoners' Dilemma think it is rational to rat no matter how much alike the two partners may be, and no matter how certain they may be that they will decide alike. Our reason is that one is better off if he rats than he would be if he didn't, since he would be ratted on or not regardless of whether he ratted. These two opinions also are one.
>
> Some have fended off the lessons of Newcomb's Problem by saying: "Let us not have, or let us not rely on, any intuitions about what is rational in goofball cases so unlike the decision problems of real life." But Prisoners' Dilemmas are deplorably common in real life. They are the most down-to-earth versions of Newcomb's Problem now available. (Lewis 1979: 240)

So if, with Lewis (and almost everyone else), you think that non-cooperation (which Lewis terms "ratting") is the rational response in a one-shot Prisoner's Dilemma, then you must also think that two-boxing is the rational strategy in NP. But then your approach to the Prisoner's Dilemma commits you to rejecting evidential decision theory. The compelling arguments for not cooperating in the Prisoner's Dilemma become compelling arguments against evidential decision theory.

Given the dialectical thrust of Lewis's conclusion it is not surprising that the most prominent discussions of his argument have primarily focused on the relation between Prisoner's Dilemma and evidential decision theory. Discussions have typically accepted Lewis's identity claim.[3] But they have tried to limit its applicability by arguing, contra Lewis, that evidential decision theory does not prescribe cooperation in the Prisoner's Dilemma.

[3] The careful analysis in Sobel (1985) makes clear that Lewis's argument does not apply to all Prisoner's Dilemmas – and in particular, it does not apply to Prisoner's Dilemmas where my beliefs about what I will do can vary independently of my beliefs about what the other player will do. But since the canonical cases that Lewis discusses are (as we will see in the next section) ones where my beliefs about my own and the other player's choices are highly interdependent, this does not count as much of a qualification.

Here is one example. Philip Pettit accepts what he terms "the Lewis result," but tries to break the tie between evidential decision theory and cooperating in Prisoner's Dilemma (Pettit 1988). He argues that Lewis's argument depends upon certain background assumptions that cannot be internalized by the participants in a way that will allow them to see cooperating as a maximizing strategy (except in vanishingly improbable circumstances). In his words:

> From an outside point of view there may be grounds for thinking of a prisoner's dilemma as a Newcomb problem, but those grounds fail to provide a basis on which participants might think of cooperating. The prisoner's dilemma, at best, is an unexploitable Newcomb problem. (Pettit 1988: 123)

As far as Pettit is concerned, therefore, the very widely held belief that non-cooperation is the rational (and dominant) strategy in the Prisoner's Dilemma is not an argument against evidential decision theory, because evidential decision theory cannot be brought to bear to argue for cooperation in the Prisoner's Dilemma in the same way that it can to argue for one-boxing in NP.

Susan Hurley also accepts the basic claim that the Prisoner's Dilemma and NP are really different faces of a single problem, while trying to undercut the connection between cooperation in Prisoner's Dilemma and evidential decision theory (Hurley 1991). She thinks that there is a real intuitive pull towards cooperation in the Prisoner's Dilemma, but denies that this intuitive pull has its source in evidential decision theory or the principle of maximizing expected utility. Hurley argues instead that the intuition in favor of cooperation in Prisoner's Dilemma is due to the appeal of what she terms cooperative reasoning. Cooperative reasoning is based on the perception that there is a collective causal power to bring about an agreed and mutually desired outcome. So, according to Hurley, people whose instinct it is to cooperate in Prisoner's Dilemmas (and the empirical literature suggests that this instinct is widespread) are really motivated by a disinclination to engage within the individualistic framework shared by both expected utility theory (evidential or causal) and dominance reasoning. It should be stressed that Hurley is neither endorsing nor rejecting cooperative reasoning. Rather, she is accounting for what she thinks of as the intuitive appeal of cooperation in the Prisoner's Dilemma. She also thinks that a similar appeal may explain intuitions in support of one-boxing in some versions of NP.

Without going into the details of the Hurley and Pettit arguments, the positions they sketch out seem inherently unstable. If Lewis's argument is sound, then the Prisoner's Dilemma is a notational variant of NP, and vice versa. As he notes in the passage quoted above, this means that there are two available strategies in the single problem that can be presented either as the Prisoner's Dilemma or as NP. One strategy can be presented either as cooperation or as one-boxing. The other can be presented either as non-cooperation or as two-boxing. So why should redescribing the strategy affect the grounds for adopting it? If evidential decision theory prescribes the strategy of one-boxing, and one-boxing is the same strategy as cooperating, then it is hard to see why evidential decision theory does not prescribe cooperating. Conversely, if one-boxing is supported by evidential decision theory while cooperating is not, then it is hard to see how one-boxing and cooperating can be at bottom the same strategy.

My argument in this chapter does not try to finesse the Lewis argument as Pettit and Hurley do. There are good reasons for thinking that Lewis is simply mistaken in his identity claim about NP and the Prisoner's Dilemma. So, looking again at the big picture, the fact that non-cooperation is (in the eyes of many) the rational strategy in the Prisoner's Dilemma is not an argument for causal decision theory. In the next section I begin by looking in more detail at Lewis's argument.

7.3 Lewis's argument

As a first step towards aligning Prisoner's Dilemma with NP, Lewis changes the description of the payoffs. His Prisoner's Dilemma is depicted in Figure 7.4, with the payoffs for A appearing first in each cell in Figure 7.4:

The changes are purely cosmetic and do not affect the structure of the game. In essence, each player receives $1,000,000 only if the other player cooperates, where cooperating is understood as not taking the $1,000. I have labeled the preference orderings as in Section 7.1 to show that the structure is identical to that of the standard Prisoner's Dilemma.

	B cooperates	B does not cooperate
A cooperates	\$1,000,000(a_2) & \$1,000,000(b_2)	\$0 (a_4) & \$1,001,000(b_2)
A does not cooperate	\$1,001,000 (a_1) & \$0 (b_2)	\$1,000 (a3) & \$1,000 (b_2)

Figure 7.4 Lewis's Prisoner's Dilemma

Lewis summarizes the situation for each player in the Prisoner's Dilemma as follows:

(1) I am offered \$1,000 – take it or leave it.
(2) I may be given an additional \$1,000,000, but whether or not this happens is causally independent of the choice I make.
(3) I will receive \$1,000,000 if and only if you do not take your \$1,000.

The causal independence in (2) is exactly mirrored in the classical Prisoner's Dilemma with payoffs in terms of years in prison, where each player's payoffs are fixed by the causally independent choice of the other player.

The point of reformulating the Prisoner's Dilemma in these terms is, of course, that it is now starting to look rather like NP. Lewis summarizes NP as follows:

(1) I am offered \$1,000 – take it or leave it.
(2) I may be given an additional \$1,000,000, but whether or not this happens is causally independent of the choice I make.
(3*) I will receive \$1,000,000 if and only if it is predicted that I do not take my \$1,000.

So, NP and the Prisoner's Dilemma differ only in the third condition. Plainly, therefore, we can show that NP and the Prisoner's Dilemma are equivalent by showing that (3) holds if and only if (3*) holds. Lewis's argument for the equivalence of the Prisoner's Dilemma and NP is essentially that (3*) entails (3). On the face of it, this would only take us halfway to the desired conclusion. But, although Lewis fails to point it out, we can use the reasoning that takes him from (3*) to (3) to make a case for the converse entailment.

Lewis's case that (3*) entails (3) starts by reformulating (3*) as (3**)

(3**) I will receive \$1,000,000 if and only if a certain potentially predictive process (which may go on before, during, or after my choice) yields an outcome which could warrant a prediction that I do not take my \$1,000.

This reformulation is legitimate, Lewis says, because "it is inessential to Newcomb's Problem that any prediction – in advance or otherwise – actually take place. It is enough that some potentially predictive process should go on and that whether I get my million is somehow made to depend on the outcome of that process" (Lewis 1979: 237).

Simulation is a very good potentially predictive process. One good way of predicting what I will do is to observe what a replica of me does in a similar predicament. So, the final step in Lewis's argument is that, when you (the other prisoner) are sufficiently like me to serve as a reliable replica, then your choice will serve as the potentially predictive process referenced in (3**) – i.e. your not taking the $1,000 will warrant a prediction that I do not take my $1,000. So, as a special case of (3**) we have:

(3) I will receive $1,000,000 if and only if you do not take your $1,000.

If (3) is a special case of (3**), which is itself a reformulation of (3*), then it seems reasonable to conclude that the Prisoner's Dilemma is a special case of NP. Moreover, in the circumstances we are considering, which is where you are sufficiently like me to serve as a reliable replica, the very same factors that make you a reliable replica of me make me a reliable replica of you. So if we have a suitable predictive process that warrants a prediction that I don't take my $1,000, then you don't take your $1,000. This seems to make NP a special case of Prisoner's Dilemma, which gives Lewis his identity claim.

7.4 The flaw in Lewis's argument

On Lewis's construal, NP and the Prisoner's Dilemma share two basic features:

(1) I am offered $1,000 – take it or leave it.
(2) I may be given an additional $1,000,000, but whether or not this happens is causally independent of the choice I make.

The two puzzles differ in their third condition. The Prisoner's Dilemma has:

(3) I will receive $1,000,000 if and only if you do not take your $1,000,

while NP has:

(3*) I will receive $1,000,000 if and only if it is predicted that I do not take my $1,000.

Labeling as follows:

(α) I will receive $1,000,000,
(β) It is predicted that I do not take my $1,000,
(γ) A certain potentially predictive process (which may go on before, during, or after my choice) yields an outcome which could warrant a prediction that I do not take my $1,000,
(δ) You do not take your $1,000,

we can represent the key steps as:

(3) $\alpha \leftrightarrow \delta$
(3*) $\alpha \leftrightarrow \beta$

Lewis argues that (3*) is equivalent to

(3**) $\alpha \leftrightarrow \gamma$

(3) is a special case of (3**) because $\delta \rightarrow \gamma$, in the scenario where you are sufficiently similar to me to count as a replica. In that scenario, we also have $\gamma \rightarrow \delta$, making (3**) a special case of (3) and hence establishing that NP and the Prisoner's Dilemma are really notational variants of a single problem.

The reasoning from (3) to (3**) and back is perfectly sound. In effect, the argument rests upon the biconditional $\delta \leftrightarrow \gamma$, which is compelling in the situation where we are sufficiently similar to be replicas of each other. So if there is a problem with the argument it must come either in the reformulation of (3*) as (3**) or in the original characterization of NP and the Prisoner's Dilemma.

It is hard to find fault with the reformulation of (3*) as (3**). What generates Newcomb's Problem is the connection between my receiving $1,000,000 and it being predicted that I won't take the $1,000. The manner of that prediction (whether it is made by a Predictor, a supernatural being, or a computer program) does not matter, nor does its timing. We just need a prediction, which can either post-date, pre-date, or be simultaneous with my choice. What matters for NP is the reliability of the prediction, and we can assume that reliability remains constant between (3*) and (3**).

Lewis's formulation of the Prisoner's Dilemma and NP in terms of (1), (2), and either (3) or (3*) respectively is a straightforward description of the payoff table, and we have already agreed that the version of the payoff table that he proposes for the Prisoner's Dilemma does not affect the structure of the game. So do we have to accept the argument and conclude that the Prisoner's Dilemma and NP are really just notational variants of each other? Not yet!

So far we have been discussing solely the structure of the game – that is to say, the payoffs and the contingencies. This leaves out a very important factor, namely the epistemic situation of the player(s). What matters in Newcomb's Problem is not simply that there be a two-way dependence between my receiving $1,000,000 and it being predicted that I not take the $1,000. That two-way dependence only generates a problem because I know that the contingency holds. So, if Lewis is correct that the Prisoner's Dilemma is an NP, then comparable knowledge is required in the Prisoner's Dilemma.

We can put this as follows, where "C_p –" is to be read as "Player p has a high degree of confidence that –". In order for a player p to be in an NP, it must be the case that

(4*) $C_p (\alpha \leftrightarrow \beta)$

Lewis's argument, therefore, depends upon showing that (4*) holds in the Prisoner's Dilemma – that each of the prisoners has a high degree of confidence that they will receive $1,000,000 if and only if they do not take the $1,000.

This really changes the structure of the argument. For one thing, Lewis needs to show not just that (3*) can be reformulated as (3**), but also that (4*) can be reformulated as

(4**) $C_p (\alpha \leftrightarrow \gamma)$

This step is not problematic. If I have a high degree of confidence in (3*) then I really ought to have a high degree of confidence in (3**), and vice versa. So it is plausible that (4*) $\leftrightarrow$ (4**). But the crucial question is how we show that (4**) holds in the Prisoner's Dilemma.

First, let's fix the starting point. If I am in a Prisoner's Dilemma then I know the payoff table. So the following holds:

(4) $C_p (\alpha \leftrightarrow \delta)$

To get Lewis's desired conclusion, which is that (4) is a special case of (4**) and (4**) a special case of (4), we can in effect repeat the earlier reasoning within the scope of the confidence operator. The key step, therefore, is

(5) $C_p (\gamma \leftrightarrow \delta)$

For (5) to hold the player must have a high degree of confidence that the other player will not take the $1,000 if and only if it is predictable that he himself will not take the $1,000. In the scenario that Lewis envisages what underwrites this high degree of confidence is the knowledge that the two players are sufficiently similar for each to serve as a simulation of the other. This is what allows me to be confident that the other player's taking her $1,000 is predictive of my taking my $1,000 – and that the predictability of my not taking my $1,000 predicts her not taking her $1,000.

This is the key point. Suppose that (5) holds for the reasons that Lewis gives. Then I am committed to a very low degree of confidence in the genuine possibility of my taking my $1,000 while the other player does not take her $1,000 – and similarly to a very low degree of confidence in the genuine possibility of my not taking my $1,000 while the other player takes her $1,000. At the limit, where I believe that the other person is a perfect replica of me, I am committed to thinking that the two scenarios just envisaged are impossible.

But if I think that two of the four available scenarios in the payoff matrix of the Prisoner's Dilemma are to all intents and purposes ruled out, then I cannot believe that I am in a Prisoner's Dilemma. In effect, what I am committed to believing is that my only two live alternatives are the upper left and bottom right scenarios in the matrix – the scenarios where we both cooperate or we both fail to cooperate. In other words, I am committed to thinking that I am in a completely different decision problem – in particular, that I am in a decision problem that is most certainly not a Prisoner's Dilemma, because it lacks the outcome scenarios that make the Prisoner's Dilemma so puzzling. So, there is an inverse correlation between (5) (which tracks my degree of confidence in the similarity between me and the other player) and my degree of confidence that I am in a Prisoner's Dilemma. Since Lewis's argument that the Prisoner's Dilemma and NP are really notational variants of a single problem rests upon (5), this means that Lewis's argument effectively undermines itself. He uses (5) to argue that the Prisoner's Dilemma is an NP, while (5) has the consequence that the player of whom (5) is true cannot believe with confidence that he is in a Prisoner's Dilemma. But any player in a Prisoner's Dilemma must believe that they are in a Prisoner's Dilemma – or rather, that they are in a game that has the structure and payoffs of a Prisoner's Dilemma. So, putting it all together, the considerations that Lewis brings to bear to show that the game he starts with is an NP equally show that the game is not a Prisoner's Dilemma.

7.5 The symmetry argument

There is a superficial similarity between the reasoning in the previous section and the so-called *symmetry argument* in favor of cooperating in a one-shot Prisoner's Dilemma. The argument was originally propounded in Davis (1977), but has failed to gain widespread acceptance (for careful discussion and representative objections, see, e.g., Bicchieri and Greene 1997). My point here rests upon symmetry considerations (as, of course, does Lewis's own argument), but is importantly different from the symmetry argument. This section explains these differences.

According to the symmetry argument, in a Prisoner's Dilemma where the two players are both rational and are in a situation of common knowledge, they can each use that knowledge to reason to a strategy of cooperation – effectively by reducing the original four possible outcomes to the two possible outcomes where both choose the same way, and then observing that each fares better in the scenario of joint cooperation than in the scenario of joint non-cooperation. The symmetry argument claims, in effect, that there are only two "live" outcomes in a standard Prisoner's Dilemma – the outcome where both cooperate and the outcome where neither cooperates. Since the outcome of joint cooperation (with a payoff for each player of $1,000,000) is clearly better for each player than the outcome of joint non-cooperation (with each player receiving $1,000), the symmetry argument states that it is rational for each player to choose to cooperate (see Davis 1977 and 1985 for more detailed expositions).

Here is a representative objection to the symmetry argument from Cristina Bicchieri and Mitchell S. Greene:

> A provisional commitment to perform a certain action is rational only if that commitment is stable under consideration of alternative possibilities, and Player 1's provisional commitment to play C [i.e. cooperate] is not stable in this way. To see this, suppose that player 1 has formed a plan to play C, and takes it that player 2 will play C as well. If player 1 now asks herself what would happen if she were to play D [i.e. not cooperate], the answer would appear to be that player 2 would continue to play C. For on a familiar construal of subjunctive conditionals, "Were A the case, then B would be the case", is true at world *w* iff, in the world most similar to *w* in which A is true, B is true as well. In the world now in question, both players play C. Because by assumption there is no causal interaction between the two players, the most similar world to *w* in which player 1 plays D (and has the same beliefs and utility function he has in *w*) is one in which Player 2 continues to play C. Since, however, (D, C) nets player 1 more than does (C, C), it follows that player 1's provisional commitment to perform C is not rational. Because the cooperative choice is not robust under consideration of alternative possibilities, both players will reach the conclusion that D is the only rational choice. (Bicchieri and Greene 1997: 236–237)

A defender of the symmetry argument certainly has ways of responding to this objection. Is it really the case that, if we start in the world of joint cooperation, the nearest world in which player 1 fails to cooperate is one in which player 2 continues to cooperate? Obviously everything depends upon the chosen similarity metric – upon how one measures similarity across possible worlds. Given that the symmetry theorist's basic claim is that there are really only two "live" possibilities, then it is easy to see that the conclusion of the symmetry argument is in effect a claim about similarity. What the symmetry argument says is that the (D, D) world, in Bicchieri and Greene's terminology, is much closer to the (C, C) world than either the (D, C) or (C, D) worlds. In fact, since the asymmetric (D, C) and (C, D) worlds are not really live possibilities at all, then the (D, C) and (C, D) worlds are not "accessible" from the (C, C) or (D, D) worlds at all (or from the actual world, for that matter).

As this brief discussion brings out, however, the symmetry argument is ultimately an argument about the structure of the Prisoner's Dilemma. If it is a sound argument then the Prisoner's Dilemma, as it is standardly construed, cannot exist. In a standard Prisoner's Dilemma there are four possible outcomes and four possible payoffs, relative to a set of starting assumptions true of both players. What the symmetry argument claims is that the starting assumptions effectively collapse those four outcomes into two. That is the real problem with the symmetry argument. The cure kills the patient, because the cooperative strategy is only a rational solution when the Prisoner's Dilemma has been transformed into a completely different game.

It should be clear how this discussion relates to my objection to Lewis's argument. As the reconstruction and discussion in the last two sections has brought out, Lewis's argument depends upon each player in the Prisoner's Dilemma thinking in a certain way about the game. In particular, the putative identity of the Prisoner's Dilemma and NP depends upon each player viewing the other player as a potentially predictive replica. As Lewis recognizes, the perceived degree of similarity and hence the perceived degree of predictability can vary.

At the limit we have a perception of complete similarity and hence both players effectively believe themselves to be in a symmetrical game. If they believe themselves to be in a symmetrical game, and satisfy the basic rationality and common knowledge requirements of the standard Prisoner's Dilemma, then they cannot believe that they are in a Prisoner's Dilemma. They must believe that they are in a game with two rather than four possible outcomes – a game where the asymmetric outcomes are not live possibilities. Of course, the perceived degree of similarity may be less than complete. In that case the players may decide that they are in a game with four possible outcomes, but where two of the outcomes are much more likely than the other two. But this game is still different from the Prisoner's Dilemma as standardly construed, which does not differentiate between the different possible outcomes. So rational players will still not believe that they are in a standard Prisoner's Dilemma.

That is my point against Lewis. A player's degree of confidence that the other player is a potentially predictive replica is inversely correlated (subject to conditions of rationality and common knowledge) with their degree of confidence that they are in a standard Prisoner's Dilemma. Moreover, rational players with common knowledge will believe that they are in a standard Prisoner's Dilemma iff they really are in a standard Prisoner's Dilemma. Therefore, each player's degree of confidence that the other player is a potentially predictive replica is inversely correlated with their really being in a standard Prisoner's Dilemma.

In sum, Lewis's argument trades on exactly the same assumptions of similarity and symmetry that we find in the symmetry argument. Both arguments are self-defeating, but they are self-defeating in different ways. The symmetry argument is *directly* self-defeating because it rests on assumptions that transform the Prisoner's Dilemma into a completely different game. Lewis's argument is *indirectly* self-defeating, because it has the same result through transforming each player's perception of the game.

7.6 The fundamental problem: parametric vs. strategic choice

Abstracting away from the details of Lewis's argument, there are good reasons for thinking that it could not possibly be sound, because NP and the Prisoner's Dilemma involve two very different types of choice. NP is standardly discussed as a problem in decision theory, which is a theory of *parametric choice*, while the Prisoner's Dilemma is a problem in game theory, which is a theory of *strategic choice*.

An agent in a parametric choice situation is choosing against a fixed background. There is variation only in one dimension – the agent's own choice. Therefore, the agent seeks to maximize expected utility relative to parameters that are set by the environment, and the rationality of her choice depends only on her preferences and probability assignments, within the context set by the parameters. Strategic choice, in contrast, involves at least one other player and there are as many dimensions of variation as there are players. The rationality of the agent's choices depends not just on her preferences and probability assignments, but also on the choices made by the other players – with the rationality of those choices also being partly determined by the agent's choices.

Game theory is the theory of strategic choice. Here is a classic characterization of a game, from Luce and Raiffa's famous textbook:

> There are *n* players each of whom is required to make one choice from a well-defined set of possible choices, and these choices are made without any knowledge as to the choices of the other players . . .Given the choices of each of the players, there is a resulting outcome which is appraised by each of the players according to his own peculiar tastes and preferences. The problem for each player is: What choice should be made in order that his partial influence over the outcome benefits him the most. He is to assume that each of the other players is similarly motivated. (Luce and Raiffa 1957: 5–6)

Plainly, the Prisoner's Dilemma satisfies this description (as one would expect, given that the Prisoner's Dilemma is the most famous example of a strategic choice problem).

But NP, as it is standardly characterized, fails to satisfy any element of Luce and Raiffa's characterization. First, the Predictor to all intents and purposes knows what the player will choose and so is not acting "without any knowledge as to the choices of the other players." In fact (and this is the second point) the Predictor does not strictly speaking choose at all. What the Predictor does is a function of what the player does, rather than being an independent variable, as would be required for a genuine strategic choice problem. Third, the Predictor does not have a preference ordering defined over the possible outcomes.

It is true that some versions of NP give the Predictor a preference ordering. Here is an ingenious suggestion from Susan Hurley:

> Think of the predictee as an intelligent Child and the predictor as a Parent who wants the Child not to be greedy on a particular occasion. Let us help ourselves explicitly to the preference orderings of each. (Of course, we're not given any such preference ordering in Newcomb's problem, which is why it is unwarranted if natural to apply cooperative reasoning.) The Child simply prefers getting more money to less. The Parent doesn't mind about whether his prediction is right or not; what he most prefers is that the Child not be greedy on this occasion, that is, that he take one box rather than two; this concern has priority over concern with saving money. But as between two situations in which the Child takes the same number of boxes, the Parent prefers the one that costs him less money. (Hurley 1994: 69–70)

It is true that these preference orderings are isomorphic to those in the Prisoner's Dilemma. Nonetheless, Hurley's attribution of a preference ordering to the Predictor/Parent does not turn NP into a Prisoner's Dilemma.

Hurley has given the Predictor a preference ordering over the possible outcomes. But that preference ordering does not determine the Predictor's choice. As in the classic NP, what the Predictor does (i.e. putting money in the opaque box or not) is a function of what the Predictor thinks that the predictee will do. If we lose the connection between the Predictor's prediction and the contents of the opaque box, then we have completely abandoned NP. So, the preference ordering that Hurley attributes to the Predictor is really a wheel spinning idly. The Predictor still does not choose, even though he now has an ordinal ranking of the different possible outcomes created by the agent's choice.

For all these reasons, then, NP and the Prisoner's Dilemma are decision problems of fundamentally different types. NP is a decision-theoretic problem of parametric choice, while Prisoner's Dilemma is a game-theoretic problem of strategic choice. It is not surprising, therefore, that Lewis's argument fails.

7.7 Conclusion

Lewis's argument that the Prisoner's Dilemma and NP are essentially notational variants of a single problem promises a fundamental contribution to debates about causal decision theory. If sound it shows that the NP is not a recondite and fanciful thought experiment, but rather an abstract structure that is realized by a vast number of social interactions. As he puts it in the passage quoted earlier, some have fended off the lessons of Newcomb's Problem by saying: "Let us not have, or let us not rely on, any intuitions about what is rational in goofball cases so unlike the decision problems of real life. But Prisoners' Dilemmas are deplorably common in real life. They are the most down-to-earth versions of Newcomb's Problem now available."

As I have brought out in this chapter, however, Lewis's argument is not sound and there remain no good reasons for thinking that the Prisoner's Dilemma is a version of NP. This is exactly what one would expect, given that NP is a problem in parametric reasoning, while the Prisoner's Dilemma is a problem in strategic reasoning. There may be good reasons for being a causal decision theorist. And there may be good reasons for two-boxing in NP. But neither causal decision theory nor one-boxing can be supported through the widely accepted rationality of not cooperating in the Prisoner's Dilemma. There remains room in logical space for non-cooperating one-boxers, and for cooperating two-boxers.

8 Counterfactuals and the Prisoner's Dilemma

Giacomo Bonanno

8.1 Introduction

In 2011 Harold Camping, president of Family Radio (a California-based Christian radio station), predicted that Rapture (the taking up into heaven of God's elect people) would take place on May 21, 2011. In light of this prediction some of his followers gave up their jobs, sold their homes and spent large sums promoting Camping's claims.[1] Did these people act rationally? Consider also the following hypothetical scenarios. Early in 2014, on the basis of a popular reading of the Mayan calendar, Ann came to believe that the world would end on December 21, 2014. She dropped out of college, withdrew all the money she had in her bank account, and decided to spend it all on traveling and enjoying herself. Was her decision rational? Bob smokes two packets of cigarettes a day; when asked if he would still smoke if he knew that he was going to get lung cancer from smoking, he answers "No"; when asked if he is worried about getting lung cancer, he says that he is not and explains that his grandfather was a heavy smoker all his life and died – cancer free – at the age of 98. Bob believes that, like his grandfather, he is immune from lung cancer. Is Bob's decision to continue smoking rational?

I will argue below that the above questions are closely related to the issue, hotly debated in the literature, whether it can be rational for the players to choose "Cooperation" in the Prisoner's Dilemma game, shown in Figure 8.1. It is a two-player, simultaneous game where each player has two strategies: "Cooperation" (denoted by *C*) and "Defection" (denoted by *D*). In each cell of the table, the first number is the utility (or payoff) of Player 1 and the second number is the utility of Player 2.

What constitutes a rational choice for a player? We take the following to be the basic definition of rationality (BDR):

[1] http://en.wikipedia.org/wiki/Harold_Camping_Rapture_prediction (accessed May 10, 2014).

A choice is rational if it is optimal given the decision-maker's preferences and beliefs. (BDR)

		Player 2	
		C	*D*
Player 1	*C*	2 , 2	0 , 3
	D	3 , 0	1 , 1

Figure 8.1 The Prisoner's Dilemma game

More precisely, we say that it is rational for the decision-maker to choose action *a* if there is no other feasible action *b* which – *according to her beliefs* – would yield an outcome that she prefers to the outcome that – again, according to her beliefs – would be a consequence of taking action *a*. According to this definition, the followers of Harold Camping did act rationally when they decided to sell everything and devote themselves to promoting Camping's claim: they believed that the world was soon coming to an end and, presumably, they viewed their proselytizing as "qualifying them for Rapture," undoubtedly an outcome that they preferred to the alternative of enduring the wrath of Judgment Day. Similarly, Ann's decision to live it up in anticipation of the end of the world predicted by the Mayan calendar qualifies as rational, as does Bob's decision to carry on smoking on the belief that – like his grandfather – he will be immune from lung cancer. Thus anybody who argues that the above decisions are *not* rational must be appealing to a stronger definition of rationality than BDR: one that denies the rationality of holding those beliefs.

When the rationality of beliefs is called into question, an asymmetry is introduced between preferences and beliefs. Concerning preferences it is a generally accepted principle that *de gustibus non est disputandum* (in matters of taste, there can be no disputes). According to this principle, there is no such thing as an irrational preference. As Rubinstein notes,

> According to the assumption of rationality in economics, the decision maker is guided by his preferences. But the assumption does not impose a limitation on the reasonableness of preferences. The preferences can be even in direct contrast with what common sense might define as the decision maker's interests. (Rubinstein 2012: 49)

For example, I cannot be judged to be irrational if I prefer an immediate benefit (e.g. from taking a drug) with known negative future consequences (e.g. from addiction) over an immediate sacrifice (e.g. by enduring pain) followed by better long-term health.[2]

In the matter of beliefs, on the other hand, it is generally thought that one *can* contend that some particular beliefs are "unreasonable" or "irrational," by appealing to such arguments as the lack of supporting evidence, the incorrect processing of relevant information, the denial of laws of Nature, etc.

Consider now the following statement by Player 1 in the Prisoner's Dilemma ("COR" stands for "correlation"):

I believe that if I play *C* then Player 2 will play *C* and that if I play *D* then Player 2 will play *D*. Thus, if I play *C* my payoff will be 2 and if I play *D* my payoff will be 1. Hence I have decided to play *C*. (COR_1)

Given the reported beliefs, Player 1's decision to play *C* is rational according to definition BDR. Thus, in order to maintain that it is not rational, one has to argue that the beliefs expressed in COR_1 violate some principle of rationality. In the literature, there are those who claim that Player 1's reported beliefs are irrational and those who claim that those beliefs can be rationally justified, for example by appealing to the symmetry of the game (see, for example, Brams 1975 and Davis 1977, 1985) or to special circumstances, such as the players being identical in some sense (e.g. they are identical twins): this has become known as the "Identicality Assumption" (this expression is used, for example, in Bicchieri and Greene 1999 and Gilboa 1999).

In order to elucidate what is involved in Player 1's belief "if I play *C* then Player 2 will play *C*, and if I play *D* then Player 2 will play *D*" we need to address the issue of the role of beliefs and conditionals in game-theoretic reasoning. In Section 8.2 we discuss the notion of a model of a game, which provides an explicit representation of beliefs and choices. After arguing that such models do not allow for an explicit discussion of rational choice, we turn in Sections 8.3–8.5 to enriched models that contain an explicit representation of subjunctive conditionals and discuss two alternative approaches: one based on belief revision and the other on objective counterfactuals. In Section 8.6 we review the few contributions in the literature that have offered a definition of rationality in strategic-form games based on an explicit appeal to counterfactuals. In Section 8.7 we discuss alternative ways of dealing with the conditionals involved in deliberation, and Section 8.8 concludes.

[2] For a criticism of the view that preferences are not subject to rational scrutiny, see chapter 10 of Hausman (2012).

8.2 Models of games: beliefs and choices

It is a widely held opinion that the notion of rationality involves the use of counterfactual reasoning. For example, Aumann (1995: 15) writes:

> [O]ne really cannot discuss rationality, or indeed decision making, without substantive conditionals and counterfactuals. Making a decision means choosing among alternatives. Thus one must consider hypothetical situations – what would happen if one did something different from what one actually does. [I]n interactive decision making – games – you must consider what other people would do if you did something different from what you actually do.

How is counterfactual reasoning incorporated in the analysis of games? The definition of strategic-form game provides only a partial description of an interactive situation. A *game in strategic form with ordinal preferences* is defined as a quintuple $G = \langle N, \{S_i\}_{i \in N}, O, z, \{\succeq_i\}_{i \in N} \rangle$, where $N = \{1,\ldots, n\}$ is a set of *players*, S_i is the set of *strategies* of (or possible choices for) player $i \in N$, O is a set of possible *outcomes*, $z: S \to O$ is a function that associates an outcome with every strategy profile $s = (s_1,\ldots, s_n) \in S = S_1 \times \ldots \times S_n$ and $\succeq_i$ is a complete and transitive binary relation on O representing player *i*'s ranking of the outcomes (the interpretation of $o \succeq_i o'$ is that player i considers outcome o to be at least as good as outcome o').[3] Games are typically

[3] Throughout this chapter we view the Prisoner's Dilemma as a strategic-form game with ordinal preferences as follows: $N = \{1, 2\}$, $S_1 = S_2 = \{C, D\}$, $O = \{o_1, o_2, o_3, o_4\}$, $z(C, C) = o_1$, $z(C, D) = o_2$, $z(D, C) = o_3$, $z(D, D) = o_4$, Player 1's ranking of the outcomes is $o_3 \succ_1 o_1 \succ_1 o_4 \succ_1 o_2$ (where $\succ$ denotes strict preference, that is, $x \succ y$ if and only if $x \succeq y$ and not $y \succeq x$) and Player 2's ranking is $o_2 \succ_2 o_1 \succ_2 o_4 \succ_2 o_3$. A preference relation $\succeq$ over the set of outcomes O can also be represented by means of an *ordinal utility function* $U: O \to \mathbb{R}$ (where $\mathbb{R}$ denotes the set of real numbers) which satisfies the property that, for any two outcomes o and o', $U(o) \geq U(o')$ if and only if $o \succeq o'$. In Figure 8.1 we have replaced each outcome with a pair of numbers, where the first is the utility of that outcome for Player 1 and the second is Player 2's utility. We take preferences over the outcomes as primitives (and utility functions merely as tools for representing those preferences). Thus we are not following the *revealed preference* approach, where observed choices are the primitives and preferences (or utility) are a derived notion:

"In revealed-preference theory, it isn't true [. . .] that Pandora chooses b rather than a because she prefers b to a. On the contrary, it is because Pandora chooses b rather than a that we say that Pandora prefers b to a, and assign b a larger utility." (Binmore 2011: 19)

Thus in the Prisoner's Dilemma game of Figure 8.1,

"Writing a larger payoff for Player 1 in the bottom-left cell of the payoff table than in the top-left cell is just another way of registering that Player 1 would choose D if she knew that Player 2 were going to choose C. [W]e must remember that Player 1 doesn't choose D because she then gets a larger payoff. Player 1 assigns a larger payoff to [the outcome associated with]

represented in *reduced form* by replacing the triple $\langle O, z, \{\succsim_i\}_{i\in N}\rangle$ with a set of *payoff functions* $\{\pi_i\}_{i\in N}$, where $\pi_i : S \to \mathbb{R}$ is any numerical function that satisfies the property that, for all s, $s' \in S$, $\pi_i(s) \geq \pi_i(s')$ if and only if $z(s) \succsim_i z(s')$, that is, if player i considers the outcome associated with s to be at least as good as the outcome associated with s'. In the following we will adopt this more succinct representation of strategic-form games (as we did in Figure 8.1).[4] Thus the definition of strategic-form game only specifies what choices each player has available and how the player ranks the possible outcomes; it is silent on what the player believes. In order to complete the description one needs to introduce the notion of *model of a game*.

Definition 1. Given a strategic-form game G, a *model of G* is a triple $\langle \Omega, \{\sigma_i\}_{i\in N}, \{\mathcal{B}_i\}_{i\in N}\rangle$ where Ω is a set of *states* and, for every player $i \in N$, $\sigma_i : \Omega \to S_i$ is a function that associates with every state $\omega \in \Omega$ a strategy $\sigma_i(\omega) \in S_i$ of player i and $\mathcal{B}_i \subseteq \Omega \times \Omega$ is a binary "doxastic" relation representing the beliefs of player i. The interpretation of $\omega\mathcal{B}_i\omega'$ is that at state ω player i considers state ω' possible. Let $\mathcal{B}_i(\omega) = \{\omega' \in \Omega : \omega\mathcal{B}_i\omega'\}$; thus $\mathcal{B}_i(\omega)$ is the set of states that player i considers possible at state ω.[5]

The functions $\{\sigma_i : \Omega \to S_i\}_{i\in N}$ give content to the players' beliefs. If $\sigma_i(\omega) = x \in S_i$ then the usual interpretation is that at state ω player i "chooses" strategy x. The exact meaning of "choosing" is not elaborated further in the literature: does it mean that player i has *actually played* x or that she *will play* x or that x is the *output of her deliberation process*? We will adopt the latter interpretation: "player i chooses x" will be taken to mean "player i has irrevocably made up her mind to play x." Subsets of Ω are called *events*. Given a state $\omega \in \Omega$ and an event $E \subseteq \Omega$, we say that *at ω player i believes E* if

(*D,C*) than to [the outcome associated with] (*C,C*) because she would choose the former if given the choice." (Binmore 2011: 27–28, with minor modifications to adapt the quotation to the notation used in Figure 8.1)

For a criticism of (various interpretations of) the notion of revealed preference, see chapter 3 of Hausman (2012); see also Rubinstein and Salant (2008).

[4] It is important to note, however, that the payoff functions are taken to be purely ordinal and one could replace π_i with any other function obtained by composing π_i with an arbitrary strictly increasing function on the set of real numbers. In the literature it is customary to impose a stronger assumption on players' preferences, namely that each player has a complete and transitive preference relation on the set of probability distributions over the set of outcomes O which satisfies the axioms of Expected Utility. For our purposes this stronger assumption is not needed.

[5] Thus the relation $\mathcal{B}_i$ can also be viewed as a function $\mathcal{B}_i : \Omega \to 2^{\Omega}$; such functions are called *possibility correspondences* in the literature. For further details the reader is referred to Battigalli and Bonanno (1999).

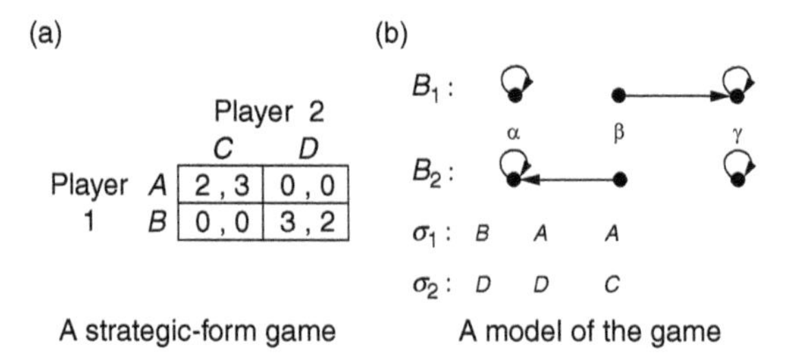

Figure 8.2 (a) A strategic-form game (b) A model of the game

and only if $\mathcal{B}_i(\omega) \subseteq E$. Part *a* of Figure 8.2 shows a strategic-form game and Part *b* a model of it (we represent a relation $\mathcal{B}$ graphically as follows: $\omega\mathcal{B}\omega'$ – or, equivalently, $\omega' \in \mathcal{B}(\omega)$ – if and only if there is an arrow from ω to ω').

State β in the model of Figure 8.2 represents the following situation: Player 1 has made up her mind to play A and Player 2 has made up his mind to play D; Player 1 erroneously believes that Player 2 has made up his mind to play C ($\mathcal{B}_1(\beta) = \{\gamma\}$ and $\sigma_2(\gamma) = C$) and Player 2 erroneously believes that Player 1 has made up her mind to play B ($\mathcal{B}_2(\beta) = \{\alpha\}$ and $\sigma_1(\alpha) = B$).

Remark 1. The model of Figure 8.2 reflects a standard assumption in the literature, namely that *a player is never uncertain about her own choice*: any uncertainty has to do with the other players' choices. This requirement is expressed formally as follows: for every $\omega' \in \mathcal{B}_i(\omega)$, $\sigma_i(\omega') = \sigma_i(\omega)$ – that is, if at state ω player i chooses strategy $x \in S_i$ ($\sigma_i(\omega) = x$) then at ω she believes that she chooses x. We shall revisit this point in Section 8.7.

Returning to the model of Part *b* of Figure 8.2, a natural question to ask is whether the players are rational at state β. Consider Player 1: according to her beliefs, the outcome is going to be the one associated with the strategy pair (A,C), with a corresponding payoff of 2 for her. In order to determine whether the decision to play A is rational, Player 1 needs to ask herself the question: "What would happen if, instead of playing A, I were to play B?" The model is silent about such counterfactual scenarios. Thus the definition of model introduced above appears to lack the resources to address the issue of rational choice.[6] Before we discuss how to enrich the definition of model

[6] It should be noted, however, that a large literature that originates in Aumann (1987) defines rationality in strategic-form games using the models described above, without enriching them with an explicit framework for counterfactuals. However, as Shin (1992: 412) notes, "If counterfactuals are not explicitly invoked, it is because the assumptions are buried implicitly in the discussion." We shall return to this point in Section 8.6.

(Sections 8.4 and 8.5), we turn, in the next section, to a brief digression on the notion of counterfactual.

8.3 Stalnaker–Lewis selection functions

There are different types of conditionals. A conditional of the form "If John received my message he will be here soon" is called an *indicative* conditional. Conditionals of the form "If I were to drop this vase, it would break" and "If we had not missed the connection, we would be at home now" are called *subjunctive* conditionals; the latter is also an example of a *counterfactual*, namely a conditional with a false antecedent (we did in fact miss the connection). It is controversial how best to classify conditionals and we will not address this issue here. We are interested in the use of conditionals in the analysis of games and thus the relevant conditionals are those that pertain to deliberation.

In the decision-theoretic and game-theoretic literature the conditionals involved in deliberation are usually called "counterfactuals," as illustrated in the quotation from Aumann (1995) in the previous section and in the following:

> [R]ational decision-making involves conditional propositions: when a person weighs a major decision, it is rational for him to ask, for each act he considers, what would happen if he performed that act. It is rational, then, for him to consider propositions of the form "If I were to do *a*, then *c* would happen." Such a proposition we shall call a counterfactual. (Gibbard and Harper 1978: 153)

With the exception of Shin (1992), Bicchieri and Greene (1999), Zambrano (2004), and Board (2006) (whose contributions are discussed in Section 8.6), the issue of counterfactual reasoning in strategic-form games has not been dealt with explicitly in the literature.[7]

We denote by $\phi > \psi$ the conditional "if ϕ were the case then ψ would be the case." In the Stalnaker–Lewis theory of conditionals (Stalnaker 1968, Lewis 1973) the formula $\phi > \psi$ has a truth value which is determined as follows: $\phi > \psi$ is true at a state ω if and only if ψ is true at all the ϕ-states that are closest (that is, most similar) to ω (a state ω' is a ϕ-state if and only if ϕ is true at ω'). While Stalnaker postulates that, for every state ω and formula ϕ,

[7] On the other hand, counterfactuals have been explored extensively in the context of dynamic games. See Bonanno (2013a) for a general discussion and relevant references.

there is a unique ϕ-state ω' that is closest to ω, Lewis allows for the possibility that there may be several such states.

The semantic representation of conditionals is done by means of a *selection function* $f: \Omega \times 2^{\Omega} \rightarrow 2^{\Omega}$ (where 2^{Ω} denotes the set of subsets of Ω) that associates with every state ω and subset $E \subseteq \Omega$ (representing a proposition) a subset $f(\omega, E) \subseteq E$ interpreted as the states in E that are closest to ω. Several restrictions are imposed on the selection function, but we will skip the details.[8]

Just as the notion of doxastic relation enables us to represent a player's beliefs without, in general, imposing any restrictions on the content of those beliefs, the notion of selection function enables us to incorporate subjunctive conditionals into a model without imposing any constraints on what ϕ-states ought to be considered most similar to a state where ϕ is not true. A comic strip shows the following dialogue between father and son:[9]

FATHER: No, you can't go.
SON: But all my friends . . .
FATHER: If all your friends jumped off a bridge, would you jump too?
SON: Oh, Jeez. . . Probably.
FATHER: What!? Why!?
SON: Because all my friends did. Think about it: which scenario is more likely? Every single friend I know – many of them levelheaded and afraid of heights – abruptly went crazy at exactly the same time . . .or the bridge is on fire?

The issue of determining what state(s) ought to be deemed closest to a given state is not a straightforward one. Usually "closeness" is interpreted in terms of a *ceteris paribus* (other things being equal) condition. However, typically *some* background conditions *must* be changed in order to evaluate a counterfactual. Consider, for example, the situation represented by state β in the model of Figure 8.2. What would be – in an appropriately enriched model – the closest state to β – call it η – where Player 1 plays B rather than A? It has been argued (we will return to this point later) that it ought to be postulated that η is a state where Player 1 has the same beliefs about Player 2's choice as in state β. Thus η would be a state where Player 1 plays B while believing that Player 2 plays C; hence at state η one of the background conditions that describe state β no longer holds, namely that Player 1 is rational and believes

[8] For example, the restriction that if $\omega \in E$ then $f(\omega, E) = \{\omega\}$.
[9] Found on the web site http://xkcd.com/1170.

herself to be rational. Alternatively, if one wants to hold this condition constant, then one must postulate that at η Player 1 believes (or at least considers it possible) that Player 2 plays D and thus one must change another background condition at β, namely her beliefs about Player 2. We will return to this issue in Section 8.6.

There is also another issue that needs to be addressed. The selection function f is usually interpreted as capturing the notion of "causality" or "objective possibility." For example, suppose that Ann is facing two faucets, one labeled "hot" and the other "cold," and she needs hot water. Suppose also that the faucets are mislabeled and Ann is unaware of this. Then it would be objectively or causally true that "if Ann turned on the faucet labeled 'cold' she would get *hot* water"; however, she could not be judged to be irrational if she expressed the belief "if I turned on the faucet labeled 'cold' I would get *cold* water" (and acted on this belief by turning on the faucet labeled 'hot'). Since what we are interested in is the issue of rational choice, objective counterfactuals do not seem to be the relevant objects to consider: *What matters is not what would in fact be the case but what the agent believes would be the case.* We shall call such beliefs *subjective counterfactuals.* How should these subjective counterfactuals be modeled? There are two options, examined in the following sections.

8.4 Subjective counterfactuals as dispositional belief revision

One construal of subjective counterfactuals is in terms of a *subjective selection function* $f_i : \Omega \times 2^\Omega \to 2^\Omega$ such that, for every $\omega \in \Omega$ and $E \subseteq \Omega$, $f_i(\omega, E) \subseteq E$. The function f_i is interpreted as expressing, at every state, player i's *initial beliefs together with her disposition to revise those beliefs under various suppositions.* Fix a state $\omega \in \Omega$ and consider the function $f_{i,\omega}: 2^\Omega \to 2^\Omega$ given by $f_{i,\omega}(E) = f_i(\omega, E)$, for every $E \subseteq \Omega$. This function gives the initial beliefs of player i at state ω (represented by the set $f_{i,\omega}(\Omega)$) as well as the set of states that player i would consider possible, at state ω, under the supposition that event $E \subseteq \Omega$ is true (represented by the set $f_{i,\omega}(E)$), for every event E. Subjective selection functions – with the implied dispositional belief revision policy – have been used extensively in the literature on dynamic games,[10] but (to the best of my knowledge) have not been used in the analysis

[10] See, for example, Arló-Costa and Bicchieri (2007), Battigalli et al. (2013), Board (2004), Bonanno (2011), Clausing (2004), Halpern (1999, 2001), Rabinowicz (2000), Stalnaker (1996). For a critical discussion of this approach, see Bonanno (2013a).

of strategic-form games, with the exception of Shin (1992) and Zambrano (2004), whose contributions are discussed in Section 8.6.

In this context, an enriched model of a strategic-form game G is a quadruple $\langle \Omega, \{\sigma_i\}_{i\in N}, \{\mathcal{B}_i\}_{i\in N}, \{f_i\}_{i\in N}\rangle$, where $\langle \Omega, \{\sigma_i\}_{i\in N}, \{\mathcal{B}_i\}_{i\in N}\rangle$ is as defined in Definition 1 and, for every player i, $f_i : \Omega\times 2^{\Omega} \to 2^{\Omega}$ is a subjective selection function satisfying the property that, for every state ω, $f_i(\omega,\Omega) = \mathcal{B}_i(\omega)$.[11] Such enriched models would be able to capture the following reasoning of Player 1 in the Prisoner's Dilemma (essentially a restatement of COR_1):

> I have chosen to play C and I believe that Player 2 has chosen to play C and thus I believe that my payoff will be 2; furthermore, I am happy with my choice of C because – under the supposition that I play D – I believe that Player 2 would play D and thus my payoff would be 1. (COR_2)

These beliefs are illustrated by state α in the following enriched model of the Prisoner's Dilemma game of Figure 8.1: $\Omega = \{\alpha, \beta\}$, $\mathcal{B}_1(\alpha) = \{\alpha\}$, $\mathcal{B}_1(\beta) = \{\beta\}$, $f_1(\alpha, \{\alpha\}) = f_1(\alpha, \Omega) = \{\alpha\}$, $f_1(\beta, \{\beta\}) = f_1(\beta, \Omega) = \{\beta\}$, $f_1(\alpha, \{\beta\}) = \{\beta\}$, $f_1(\beta, \{\alpha\}) = \{\alpha\}$, $\sigma_1(\alpha) = C$, $\sigma_1(\beta) = D$, $\sigma_2(\alpha) = C$, $\sigma_2(\beta) = D$ (we have omitted the beliefs of Player 2). At state α Player 1 believes that she is playing C and Player 2 is playing C ($\mathcal{B}_1(\alpha) = \{\alpha\}$ and $\sigma_1(\alpha) = C$ and $\sigma_2(\alpha) = C$); furthermore the proposition "Player 1 plays D" is represented by the event $\{\beta\}$ (β is the only state where Player 1 plays D) and thus, since $f_1(\alpha, \{\beta\}) = \{\beta\}$ and $\sigma_2(\beta) = D$, Player 1 believes that – under the supposition that she plays D –Player 2 plays D and thus her own payoff would be 1.

Are the beliefs expressed in COR_2 compatible with rationality? The principles of "rational" belief revision, that are captured by the properties listed in footnote 11, are principles of logical coherence of dispositional beliefs[12] and, in general, do not impose any constraints on the content of a counterfactual belief. Thus the above beliefs of Player 1 *could* be rational beliefs, in the sense that they do not violate logical principles or principles of coherence. Those who claim that the beliefs expressed in COR_2 are irrational appeal to the argument that they imply a belief by Player 1 that her "switching" from C to

[11] Alternatively, one could remove the initial beliefs $\{\mathcal{B}_i\}_{i\in N}$ from the definition of an extended model and recover them from the function f_i by taking $f_i(\omega, \Omega)$ to be the set of states that player i initially considers possible at state ω. There are further consistency properties that are usually imposed: (1) if $E \neq \varnothing$ then $f_i(\omega, E) \neq \varnothing$, (2) if $\mathcal{B}_i(\omega) \cap E \neq \varnothing$ then $f_i(\omega, E) = \mathcal{B}_i(\omega) \cap E$, and (3) if $E \subseteq F$ and $f_i(\omega, F) \cap E \neq \varnothing$ then $f_i(\omega, E) = f_i(\omega, F) \cap E$. For a more detailed discussion, see Bonanno (2013a).

[12] The principles that were introduced by Alchourrón et al. (1985), which pioneered the vast literature on the so-called AGM theory of belief revision.

D causes Player 2 to change her decision from *C* to *D*, while such a causal effect is ruled out by the fact that each player is making her choice in ignorance of the choice made by the other player (the choices are made "simultaneously"). For example, Harper (1988: 25) claims that "a causal independence assumption is part of the idealization built into the normal form" and Stalnaker (1996: 138) writes: "[I]n a strategic form game, the assumption is that the strategies are chosen independently, which means that the choices made by one player cannot influence the beliefs or the actions of the other players." One can express this point of view by imposing the following restriction on beliefs: In an enriched model of a game, if at state ω player i considers it possible that his opponent is choosing any one of the strategies $w_1, \ldots, w_m$, then the following must be true for every strategy x of player i: under the supposition that she plays x, player i continues to consider it possible that his opponent is choosing any one of the strategies $w_1, \ldots, w_m$ and no other strategies.

This condition can be expressed more succinctly as follows. Given a state ω and a player i we denote by $\sigma_{-i}(\omega) = (\sigma_1(\omega), \ldots, \sigma_{i-1}(\omega), \sigma_{i+1}(\omega), \ldots, \sigma_n(\omega))$ the strategies chosen at ω by the players other than i; furthermore, for every strategy x of player i, let $[x]$ denote the event that (that is, the set of states at which) player i plays x. Then the above restriction on beliefs can be written as follows ("IND" stands for "independence" and "subj" for "subjective"):

For every state ω and for every $x \in S_i$,

$$\bigcup_{\omega' \in \mathcal{B}_i(\omega)} \{\sigma_{-i}\{\omega'\}\} = \bigcup_{\omega' \in f_i(\omega, [x])} \{\sigma_{-i}\{\omega'\}\} \qquad (IND_1^{subj})$$

The beliefs expressed in COR_2 violate condition IND_1^{subj}. Should IND_1^{subj} be viewed as a necessary condition for rational beliefs? This question will be addressed in Section 8.8.

8.5 Subjective counterfactuals as beliefs about causality

The usual argument in support of the thesis that, for the Prisoner's Dilemma, Player 1's reasoning expressed in COR_1 is fallacious is that even if (e.g. because of symmetry or because of the "identicality assumption") one agrees that the outcome must be one of the two on the diagonal, the off-diagonal outcomes are nevertheless causally possible. Thus one must distinguish between *causal* (or objective) possibility and *doxastic* (or subjective) possibility and in the process of rational decision-making one has to consider the relevant causal possibilities, even if they are ruled out as doxastically impossible. This is where objective counterfactuals become relevant. This line of reasoning is at the core of causal decision theory.[13]

According to this point of view, subjective counterfactuals should be interpreted in terms of the composition of a belief relation $\mathcal{B}_i$ with an objective counterfactual selection function f: $\Omega \times 2^{\Omega} \to 2^{\Omega}$. In this approach, an enriched model of a strategic-form game G is a quadruple $\langle \Omega, \{\sigma_i\}_{i \in N}, \{\mathcal{B}_i\}_{i \in N}, f \rangle$, where $\langle \Omega, \{\sigma_i\}_{i \in N}, \{\mathcal{B}_i\}_{i \in N} \rangle$ is as defined in Definition 1 and $f: \Omega \times 2^{\Omega} \to 2^{\Omega}$ is an objective selection function. In this context, $f(\omega, E)$ is the set of states in E that would be "causally true" (or objectively true) at state ω if E were the case, while $\bigcup_{\omega' \in \mathcal{B}_i(\omega)} f(\omega', E)$ is the set of states in E that – according to player i's beliefs at state ω – could be "causally true" if E were the case.

As noted in the previous section, from the point of view of judging the rationality of a choice, what matters is not the "true" causal effect of that choice but what the agent *believes* to be the causal effect of her choice, as illustrated in the example of Section 8.3 concerning the mislabeled faucets. As another example, consider the case of a player who believes to be engaged – as Player 1 – in a Prisoner's Dilemma game, while in fact Player 2 is a computer that will receive as input Player 1's choice and has been programmed to mirror that choice. In this case, in terms of objective counterfactuals, there is perfect correlation between the choices of the two players, so that the best choice of Player 1 would be to play C. However, Player 1 may rationally play D if she believes that (1) Player 2 will play D and (2) if she were to play C then Player 2 would still play D.

Causal independence, at a state ω, between the choice of player i and the choices of her opponents would be expressed by the following restriction on the objective selection function (recall that $\sigma_{-i}(\omega) = (\sigma_1(\omega), \ldots, \sigma_{i-1}(\omega), \sigma_{i+1}(\omega), \ldots, \sigma_n(\omega))$ is the profile of strategies chosen at ω by i's opponents and that, for $x \in S_i$, $[x]$ denotes the event that – that is, the set of states where – player i chooses strategy x; "obj" stands for "objective"):

For every strategy x of player i, if $\omega' \in f(\omega, [x])$, then $\sigma_{-i}(\omega') = \sigma_{-i}(\omega)$. (IND^{obj})

However, as noted above, what matters is not whether IND^{obj} holds at state ω but whether player i *believes* that IND^{obj} holds. Hence the following, subjective, version of independence is the relevant condition:

[13] There are various formulations of causal decision theory: see Gibbard and Harper (1978), Lewis (1981), Skyrms (1982), and Sobel (1986). For an overview, see Weirich (2008).

For every strategy x of player i and for every $\omega' \in \mathcal{B}_i(\omega)$, if $\omega'' \in f(\omega', [x])$ then $\sigma_{-i}(\omega'') = \sigma_{-i}(\omega')$. $\left(IND_2^{subj}\right)$

It is straightforward to check that condition IND_2^{subj} implies condition IND_1^{subj} if one defines $f_i(\omega, E) = \bigcup_{\omega' \in \mathcal{B}_i(\omega)} f(\omega', E)$, for every event E; indeed, a slightly weaker version of IND_2^{subj} is equivalent to IND_1^{subj}.[14]

We conclude that, since a player may hold erroneous beliefs about the causal effects of her own choices and what matters for rational choice is what the player believes rather than what is "objectively true," there is no relevant conceptual difference between the objective approach discussed in this section and the subjective approach discussed in the previous section.[15]

8.6 Rationality of choice: Discussion of the literature

We are yet to provide a precise definition of rationality in strategic-form games. With the few exceptions described below, there has been no formal discussion of the role of counterfactuals in the analysis of strategic-form games. Aumann (1987) was the first to use the notion of epistemic[16] model of a strategic-form game. His definition of rationality, which we will state in terms of beliefs (rather than the more restrictive notion of knowledge) and call Aumann-rationality, is as follows. Recall that, given a state ω in a model of a game and a player i, $\sigma_i(\omega)$ denotes the strategy chosen by player i at state ω, while the profile of strategies chosen by the other players is denoted by $\sigma_{-i}(\omega) = (\sigma_1(\omega), \ldots, \sigma_{i-1}(\omega), \sigma_{i+1}(\omega), \ldots, \sigma_n(\omega))$.

Definition 2. Consider a model of a strategic form game (see Definition 1), a state ω and a player i. Player i's choice at state ω is *Aumann-rational* if there is no other strategy s_i of player i such that $\pi_i(s_i, \sigma_{-i}(\omega')) > \pi_i(\sigma_i(\omega), \sigma_{-i}(\omega'))$

[14] IND_2^{subj} implies that $\bigcup_{\omega'' \in \bigcup_{\omega' \in \mathcal{B}_i(\omega)} f(\omega', [x])} \{\sigma_{-i}\{\omega''\}\} = \bigcup_{\omega' \in \mathcal{B}_i(\omega)} \{\sigma_{-i}\{\omega'\}\}$ which coincides with IND_1^{subj} if one takes $f_i(\omega, [x]) = \bigcup_{\omega' \in \mathcal{B}_i(\omega)} f(\omega', [x])$.

[15] Although in strategic-form games the two approaches can be considered to be equivalent, this is not so for dynamic games, where the "objective" approach may be too restrictive. This point is discussed in Bonanno (2013a).

[16] The models used by Aumann (1987, 1995) make use of knowledge, that is, of necessarily correct beliefs. We refer to these models as epistemic, reserving the term "doxastic" for models that use the more general notion of belief, which allows for the possibility of error. The models discussed in this chapter are the more general doxastic models.

for every $\omega' \in \mathcal{B}_i(\omega)$.[17] That is, player i's choice is rational if it is not the case that player i believes that another strategy of hers is strictly better than the chosen strategy.

The above definition is weaker than the definition used in Aumann (1987), since – for simplicity – we have restricted attention to ordinal payoffs and qualitative (that is, non-probabilistic) beliefs.[18] However, *the essential feature of this definition is that it evaluates counterfactual strategies of player i keeping the beliefs of player i constant.* Hence implicit in this definition of rationality is either a theory of subjective counterfactuals that assumes condition IND_1^{subj} or an objective theory of counterfactuals that assumes condition IND_2^{subj}. The only attempts (that I am aware of) to bring the relevant counterfactuals to the surface are Shin (1992), Bicchieri and Greene (1999), Zambrano (2004), and Board (2006).

Shin (1992) develops a framework which is very similar to one based on subjective selection functions (as described in Section 8.4). For each player i in a strategic-form game Shin defines a "subjective state space" Ω_i. A point in this space specifies a belief of player i about his own choice and the choices of the other players. Such belief assigns probability 1 to player i's own choice (that is, player i is assumed to know his own choice). Shin then defines a metric on this space as follows. Let ω be a state where player i attaches probability 1 to his own choice, call it A, and has beliefs represented by a probability distribution P on the strategies of his opponents; the closest state to ω where player i chooses a different strategy, say B, is a state ω' where player i attaches probability 1 to B and has the same probability distribution P over the strategies of his opponents that he has at ω. This metric allows player i to evaluate the counterfactual "if I chose B then my payoff would be x." Thus Shin imposes *as an axiom* the requirement that player i should hold the same beliefs about the other players' choices when contemplating a "deviation" from his actual choice. This assumption corresponds to requirement IND_1^{subj}. Not surprisingly, his main result is that a player is rational with respect to this metric if and only if she is Aumann-rational.

[17] Recall that $\mathcal{B}_i(\omega)$ is the set of states that player i considers possible at state ω; recall also the assumption that $\sigma_i(\bullet)$ is constant on $\mathcal{B}_i(\omega)$, that is, for every $\omega' \in \mathcal{B}_i(\omega)$, $\sigma_i(\omega') = \sigma_i(\omega)$.

[18] When payoffs are taken to be von Neumann–Morgenstern payoffs and the beliefs of player i at state ω are represented by a probability distribution $\mathbf{p}_{i,\omega} : \Omega \to [0, 1]$ (assuming that Ω is a finite set) whose support coincides with $\mathcal{B}_i(\omega)$ (that is, $\mathbf{p}_{i,\omega}(\omega') > 0$ if and only if $\omega' \in \mathcal{B}_i(\omega)$) then the choice of player i at state ω is defined to be rational if and only if it maximizes player i's expected payoff at state ω, that is, if and only if there is no strategy s_i of player i such that $\sum_{\omega' \in \mathcal{B}_i(\omega)} \mathbf{p}_{i,\omega}(\omega')\, \pi_i(s_i, \sigma_{-i}(\omega')) > \sum_{\omega' \in \mathcal{B}_i(\omega)} \mathbf{p}_{i,\omega}(\omega')\, \pi_i(\sigma_i(\omega), \sigma_{-i}(\omega'))$.

Zambrano's (2004) approach is a mixture of objective and subjective counterfactuals. His analysis is restricted to two-player strategic-form games. First of all, he defines a subjective selection function for player i, $f_i : \Omega \times S_i \rightarrow \Omega$, which follows Stalnaker (1968) in assuming that, for every hypothesis and every state ω, there is a *unique* world closest to ω where that hypothesis is satisfied; furthermore, the hypotheses consist of the possible strategies of player i (the set of strategies S_i), rather than events. He interprets $f_i(\omega, s_i) = \omega'$ as follows: "state ω' is the state closest to ω, according to player i, in which player i deviates from the strategy prescribed by ω and, instead, plays s_i" (p. 5). He then imposes the requirement that "player i is the *only* one that deviates from $\sigma(\omega)$ in $f_i(\omega, s_i)$, that is, $\sigma_j(f_i(\omega, s_i)) = \sigma_j(\omega)$" (Condition F2, p. 5; j denotes the other player). This appears to be in the spirit of the objective causal independence assumption IND_2^{obj}. However, Zambrano does not make use of this requirement, because he focuses on the beliefs of player i at the state $f_i(\omega, s_i)$ and uses *these* beliefs to evaluate *both* the original strategy $\sigma_i(\omega)$ *and* the new strategy s_i. He introduces the following definition of rationality:

> player i is W-rational [at state ω] if there is no deviation $s_i \neq \sigma_i(\omega)$ such that strategy s_i is preferred to $\sigma_i(\omega)$ given the belief that player i holds *at the state closest to ω in which i deviates to s_i*. The interpretation is that the rationality of choosing strategy $\sigma_i(\omega)$ at state ω against a deviation $s_i \neq \sigma_i(\omega)$ is determined with respect to beliefs that arise at the closest state to ω in which s_i is actually chosen, that is, with respect to beliefs at $f_i(\omega, s_i)$. (Zambrano 2004, p. 6).

Expressed in terms of our qualitative approach, player i is W-rational at state ω if there is no strategy s_i of player i such that $\pi_i(s_i, \sigma_{-i}(\omega')) > \pi_i(\sigma_i(\omega), \sigma_{-i}(\omega'))$ for every $\omega' \in \mathcal{B}_i(f_i(\omega, s_i))$. Hence, unlike Aumann-rationality (Definition 2), the quantification is over $\mathcal{B}_i(f_i(\omega, s_i))$ rather than over $\mathcal{B}_i(\omega)$.[19] *The definition of W-rationality thus disregards the beliefs of player i at state ω and focuses instead on the beliefs that player i would have if she changed her strategy.* Since, in general, those hypothetical beliefs can be different from the initial beliefs at state ω, there is no connection between

[19] Zambrano uses probabilistic beliefs: for every $\omega \in \Omega$, $\mathbf{p}_{i,\omega} : \Omega \rightarrow [0, 1]$ is a probability distribution over Ω that represents the beliefs of player i at state ω. Our set $\mathcal{B}_i(\omega)$ corresponds to the support of $\mathbf{p}_{i,\omega}$. Zambrano's definition is as follows: player i is W-rational at state ω if there is no strategy s_i of player i such that $\sum_{\omega' \in \Omega} \mathbf{p}_{i, f_i(\omega, s_i)}(\omega')\, \pi_i(s_i, \sigma_j(\omega')) > \sum_{\omega' \in \Omega} \mathbf{p}_{i, f_i(\omega, s_i)}(\omega')\, \pi_i(\sigma_i(\omega), \sigma_j(\omega'))$.

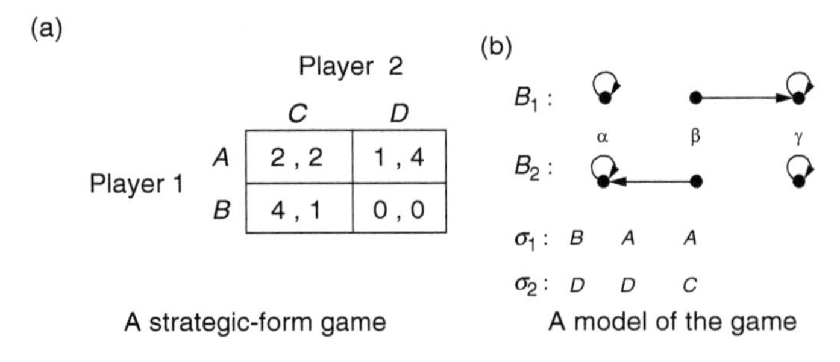

Figure 8.3 (a) A strategic-form game (b) A model of the game

W-rationality and Aumann-rationality. For example, consider the game shown in Part *a* of Figure 8.3 and the model shown in Part *b*.

Let the subjective selection function of Player 1 be given by $f_1(\alpha, B) = f_1(\beta, B) = \alpha$ and $f_1(\alpha, A) = f_1(\beta, A) = \beta$. Consider state α where the play is (B,D) and both players get a payoff of 0. Player 1 is W-rational at state α (where she chooses B and believes that Player 2 chooses D) because if she were to play A (state β) then she would believe that Player 2 played C and – given these beliefs – playing B is better than playing A. However, Player 1 is not Aumann-rational at state α, because the notion of Aumann-rationality uses the beliefs of Player 1 at state α to compare A to B (while the notion of W-rationality uses the beliefs at state β).

Zambrano then shows (indirectly, through the implications of common knowledge of rationality) that W-rationality coincides with Aumann-rationality if one adds the following restriction to the subjective selection function f_i: for every state ω and every strategy $s_i \in S_i$, at the closest state to ω where player i plays strategy s_i, the beliefs of player i concerning the strategy chosen by the other player (player j) are the same as at state ω.[20] This is in the spirit of condition IND_1^{subj}.

Board (2006) uses objective counterfactuals as defined by Stalnaker (1968) (for every hypothesis and every state ω, there is a *unique* world closest to ω where that hypothesis is satisfied). Like Zambrano, Board takes as possible hypotheses the individual strategies of the players: he introduces an *objective* selection function $f : \Omega \times \bigcup_{i \in N} S_i \to \Omega$, that specifies – for every state ω, every player i and every strategy $s_i \in S_i$ of player i – the unique world $f(\omega, s_i) \in \Omega$ closest to ω where player i chooses s_i. Recall that $\sigma_i(\omega)$ denotes the strategy chosen by player i at state ω. In accordance with

[20] Given that Zambrano postulates probabilistic beliefs, he expresses this condition as follows: $\text{marg}_{S_j} p_{i,\omega}(\bullet) = \text{marg}_{S_j} p_{i,f_i(\omega,s_i)}(\bullet)$.

Stalnaker's theory of counterfactuals, Board assumes that $f(\omega, \sigma_i(\omega)) = \omega$, that is, the closest state to ω where player i chooses the strategy that he chooses at ω is ω itself. On the other hand, if $s_i \neq \sigma_i(\omega)$ and $f(\omega, s_i) = \omega'$ then it is necessarily the case that $\omega' \neq \omega$, since it must be that $\sigma_i(\omega') = s_i$. What does player i believe at state ω about the choices of the other players? As before, let $\mathcal{B}_i$ be the belief relation of player i and $\mathcal{B}_i(\omega) = \{\omega' \in \Omega : \omega\mathcal{B}_i\omega'\}$ the belief set of player i at state ω. We denote by $S_{-i} = S_1 \times \ldots \times S_{i-1} \times S_{i+1} \times \ldots \times S_n$ the set of strategy profiles for the players other than i. Then the set of strategy profiles of the opponents that player i considers possible at state ω, if she plays her chosen strategy $\sigma_i(\omega)$, is $\bigcup_{\omega' \in \mathcal{B}_i(\omega)} \{\sigma_{-i}(\omega')\} = \{s_{-i} \in S_{-i} : s_{-i} = \sigma_{-i}(\omega') \text{ for some } \omega' \in \mathcal{B}_i(\omega)\}$. On the other hand, what are her beliefs – at state ω – about the strategy profiles of her opponents if she were to choose a strategy $s_i \neq \sigma_i(\omega)$? For every state ω' that she deems possible at state ω (that is, for every $\omega' \in \mathcal{B}_i(\omega)$) she considers the closest state to ω' where she plays s_i, namely $f(\omega', s_i)$, and looks at the choices made by her opponents at state $f(\omega', s_i)$.[21] Thus the set of strategy profiles of the opponents that player i would consider possible at state ω, if she were to play a strategy $s_i \neq \sigma_i(\omega)$, is $\bigcup_{\omega' \in \mathcal{B}_i(\omega)} \{\sigma_{-i}(f(\omega', s_i))\}$.[22] Note that, in general, there is no relationship between the sets $\bigcup_{\omega' \in \mathcal{B}_i(\omega)} \{\sigma_{-i}(f(\omega', s_i))\}$ and $\bigcup_{\omega' \in \mathcal{B}_i(\omega)} \{\sigma_{-i}(\omega')\}$; indeed, these two sets might even be disjoint.

Board defines player i to be *causally rational* at state ω (where she chooses strategy $\sigma_i(\omega)$) if it is not the case that she believes, at state ω, that there is another strategy $s_i \in S_i$ which would yield a higher payoff than $\sigma_i(\omega)$. His definition is expressed in terms of expected payoff maximization.[23] Since, in general, the two sets $\bigcup_{\omega' \in \mathcal{B}_i(\omega)} \{\sigma_{-i}(f(\omega', s_i))\}$ and $\bigcup_{\omega' \in \mathcal{B}_i(\omega)} \{\sigma_{-i}(\omega')\}$ might be

[21] Recall the assumption that a player always knows her chosen strategy, that is, for every $\omega' \in \mathcal{B}_i(\omega)$, $\sigma_i(\omega') = \sigma_i(\omega)$ and thus – since we are considering a strategy $s_i \neq \sigma_i(\omega)$ – it must be the case that $f(\omega', s_i) \neq \omega'$.

[22] This set can also be written as $\bigcup_{\omega'' \in \bigcup_{\omega' \in \mathcal{B}_i(\omega)} f(\omega', s_i)} \{\sigma_{-i}(\omega'')\}$.

[23] Like Zambrano, Board assumes that payoffs are von Neumann–Morgenstern payoffs and beliefs are probabilistic: for every $\omega \in \Omega$, $\mathbf{p}_{i,\omega}$ is a probability distribution with support $\mathcal{B}_i(\omega)$ that represents the probabilistic beliefs of player i at state ω. Board defines player i to be causally rational at state ω if there is no strategy s_i that would yield a higher expected payoff if chosen instead of $\sigma_i(\omega)$, that is, if there is no $s_i \in S_i$ such that $\sum_{\omega' \in \mathcal{B}_i(\omega)} \mathbf{p}_{i,\omega}(\omega')\, \pi_i(s_i, \sigma_{-i}(f(\omega', s_i))) > \sum_{\omega' \in \mathcal{B}_i(\omega)} \mathbf{p}_{i,\omega}(\omega')\, \pi_i(\sigma_i(\omega), \sigma_{-i}(\omega'))$. There is no clear qualitative counterpart to this definition, because of the lack of any constraints that

disjoint, causal rationality is consistent with each player choosing Cooperation in the Prisoner's Dilemma. To see this, consider the following partial model of the Prisoner's Dilemma game of Figure 8.1, where, for the sake of brevity, we only specify the beliefs of Player 1 and the objective counterfactuals concerning the strategies of Player 1 and, furthermore, in order to avoid ambiguity, we denote the strategies of Player 1 by C and D and the strategies of Player 2 by c and d: $\Omega = \{\alpha, \beta, \gamma, \delta\}$, $\sigma_1(\alpha) = \sigma_1(\beta) = C$, $\sigma_1(\gamma) = \sigma_1(\delta) = D$, $\sigma_2(\alpha) = \sigma_2(\beta) = \sigma_2(\delta) = c$, $\sigma_2(\gamma) = d$, $\mathcal{B}_1(\alpha) = \mathcal{B}_1(\beta) = \{\beta\}$, $\mathcal{B}_1(\gamma) = \{\gamma\}$, $\mathcal{B}_1(\delta) = \{\delta\}$, $f(\alpha, C) = f(\gamma, C) = f(\delta, C) = \alpha$, $f(\beta, C) = \beta$, $f(\alpha, D) = \delta$, $f(\beta, D) = f(\gamma, D) = \gamma$ and $f(\delta, D) = \delta$. Then at state α Player 1 is causally rational: she chooses C and believes that her payoff will be 2 (because she believes that Player 2 has chosen c: $\mathcal{B}_1(\alpha) = \{\beta\}$ and $\sigma_2(\beta) = c$) and she also believes that if she were to play D then Player 2 would play d ($\mathcal{B}_1(\alpha) = \{\beta\}$, $f(\beta, D) = \gamma$ and $\sigma_2(\beta) = d$) and thus her payoff would be 1. Note that at state α Player 1 has incorrect beliefs about what would happen if she played D: since $f(\alpha, D) = \delta$ and $\sigma_2(\delta) = c$, the "objective truth" is that if Player 1 were to play D then Player 2 would still play c, however Player 1 believes that Player 2 would play d. Note that state α in this model provides a formal representation of the reasoning expressed in COR_1. Board's main result is that a necessary and sufficient condition for causal rationality to coincide with Aumann-rationality is the IND_2^{subj} condition of Section 8.5.[24]

Bicchieri and Greene's (1999) aim is to clarify the implications of the "identicality assumption" in the Prisoner's Dilemma game. They enrich the definition of a model of a game (Definition 1) by adding a binary relation $\mathcal{C} \subseteq \Omega \times \Omega$ of "nomic accessibility," interpreting $\omega \mathcal{C} \omega'$ as "ω' is causally possible relative to ω" in the sense that "everything that occurs at ω' is consistent with the laws of nature that hold at ω" (p. 180). After discussing at length the difference between doxastic possibility (represented by the relations $\mathcal{B}_i$, $i \in N$) and causal possibility (in the spirit of causal decision theory), they raise the question whether it is possible to construe a situation

relate $\underset{\omega' \in \mathcal{B}_i(\omega)}{\cup} \{\sigma_{-i}(f(\omega', s_i))\}$ to $\underset{\omega' \in \mathcal{B}_i(\omega)}{\cup} \{\sigma_{-i}(\omega')\}$. Board (2006: 16) makes this point as follows: "Since each state describes what each player does as well as what her opponents do, the player will change the state if she changes her choice. There is no guarantee that her opponents will do the same in the new state as they did in the original state."

[24] Board presents this as an objective condition on the selection function (if $\omega' = f(\omega, s_i)$ then $\sigma_{-i}(\omega') = \sigma_{-i}(\omega)$) assumed to hold at every state (and thus imposed as an axiom), but then acknowledges (p. 12) that "it is players' beliefs in causal independence rather than causal independence itself that drives the result."

in which it is causally necessary that the choices of the two players in the Prisoner's Dilemma are the same, while their actions are nonetheless causally independent. They suggest that the answer is positive: one could construct an agentive analog of the Einstein–Podolsky–Rosen phenomenon in quantum mechanics (p. 184). They conclude that there may indeed be a coherent nomic interpretation of the identicality assumption, but such interpretation may be controversial.

In the next section we discuss the issue of whether subjunctive conditionals or counterfactuals – as captured by (subjective or objective) selection functions – are indeed a necessary, or even desirable, tool for the analysis of rational choice.

8.7 Conditionals of deliberation and pre-choice beliefs

A common feature of all the epistemic/doxastic models of games used in the literature is the assumption that if a player chooses a particular action at state ω then she knows, at state ω, that she chooses that action. This approach thus requires the use of either objective or subjective counterfactuals in order to represent a player's beliefs about the consequences of taking alternative actions. However, several authors have maintained that *it is the essence of deliberation that one cannot reason towards a choice if one already knows what that choice will be.* For instance, Shackle (1958: 21) remarks that if an agent could predict the option he will choose, his decision problem would be "empty," Ginet (1962: 50) claims that "it is conceptually impossible for a person to know what a decision of his is going to be before he makes it," Goldman (1970: 194) writes that "deliberation implies some doubt as to whether the act will be done," Spohn (1977: 114) states the principle that "any adequate quantitative decision model must not explicitly or implicitly contain any subjective probabilities for acts" (and later [Spohn, 2012: 109] writes that "the decision model must not impute to the agent any cognitive or doxastic assessment of his own actions"), Levi (1986: 65) states that "the deliberating agent cannot, before choice, predict how he will choose" and coins the phrase "deliberation crowds out prediction" (Levi 1997: 81).[25]

[25] Similar observations can be found in Schick (1979), Gilboa (1999), Kadane and Seidenfeld (1999); for a discussion and further references, see Ledwig (2005). It should be noted, however, that this view has been criticized by several authors: see, for example, Joyce (2002), Rabinowicz (2002), and Peterson (2006); Luce (1959) also claimed that it sometimes makes sense to assign probabilities to one's own choices.

Deliberation involves reasoning along the following lines: "If I take action a, then the outcome will be x and if I take action b, then the outcome will be y." Indeed, it has been argued (DeRose 2010) that the appropriate conditionals for deliberation are *indicative* conditionals, rather than subjunctive conditionals. If I say: "If I had left the office at 4 pm I would not have been stuck in traffic," I convey the information that – as a matter of fact – I did not leave the office at 4 pm and thus I am uttering a counterfactual conditional, namely one which has a false antecedent (such a statement would not make sense if uttered before 4 pm). On the other hand, if I say: "If I leave the office at 4 pm I will not be stuck in traffic," I am uttering what is normally called an indicative conditional and I am conveying the information that I am evaluating the consequences of a possible future action (such a statement would not make sense if uttered after 4 pm). Concerning the latter conditional, is there a difference between the indicative mood and the subjunctive mood? If I said (before 4 pm): "If I were to leave the office at 4 pm I would not be stuck in traffic," would I be conveying the same information as with the previous indicative conditional? On this point there does not seem to be a clear consensus in the literature. I agree with DeRose's claim that the subjunctive mood conveys different information relative to the indicative mood: its role is to

> call attention to the possibility that the antecedent is (or will be) false, where one reason one might have for calling attention to the possibility that the antecedent is (or will be) false is that it is quite likely that it is (or will be) false. (DeRose 2010: 10)

The indicative conditional signals that the decision whether to leave the office at 4 pm is still "open," while the subjunctive conditional intimates that the speaker is somehow ruling out that option: for example, he has made a tentative or firm decision not to leave at 4 pm.

In light of the above discussion it would be desirable to model a player's *deliberation-stage* (or *pre-choice*) beliefs, where the player considers the consequences of all her actions, *without predicting her subsequent decision*. If a state encodes the player's actual choice, then that choice can be judged to be rational or irrational by relating it to the player's pre-choice beliefs. Hence, if one follows this approach, it becomes possible for a player to have the same beliefs in two different states, ω and ω', and be labeled as rational at state ω and irrational at state ω', because the action she ends up taking at state ω is optimal given those beliefs, while the action she ends up taking at state ω' is not optimal given those same beliefs.

A potential objection to this view arises in dynamic games where a player chooses more than once along a given play of the game. Consider a situation where at time t_1 player i faces a choice and knows that she might be called upon to make a second choice at a later time t_2. The view outlined above requires player i to have "open" beliefs about her choice at time t_1 but also allows her to have beliefs about what choice she will make at the later time t_2. Is this problematic? Several authors have maintained that there is no inconsistency between the principle that one should not attribute to a player beliefs about her current choice and the claim that, on the other hand, one can attribute to the player beliefs about her later choices. For example, Gilboa writes:

> [W]e are generally happier with a model in which one cannot be said to have beliefs about (let alone knowledge of) one's own choice while making this choice. [O]ne may legitimately ask: Can you truly claim you have no beliefs about your own future choices? Can you honestly contend you do not believe – or even know – that you will not choose to jump out of the window? [T]he answer to these questions is probably a resounding "No." But the emphasis should be on timing: when one considers one's choice tomorrow, one may indeed be quite sure that one will not decide to jump out of the window. However, a future decision should actually be viewed as a decision by a different "agent" of the same decision maker. [. . .] It is only at the time of choice, within an "atom of decision," that we wish to preclude beliefs about it." (Gilboa 1999: 171 –172)

In a similar vein, Levi (1997: 81) writes that "agent X may coherently assign unconditional credal probabilities to hypotheses as to what he will do when some future opportunity for choice arises. Such probability judgments can have no meaningful role, however, when the opportunity of choice becomes the current one." Similarly, Spohn (1999: 44–45) maintains that in the case of sequential decision-making, the decision-maker can ascribe subjective probabilities to his future – but not to his present – actions. We share the point of view expressed by these authors. If a player moves sequentially at times t_1 and t_2, with $t_1 < t_2$, then at time t_1 she has full control over her immediate choices (those available at t_1) but not over her later choices (those available at t_2). The agent can predict – or form an intention about – her future behavior at time t_2, but she cannot irrevocably decide it at time t_1, just as she can predict – but not decide – how other individuals will behave after her current choice.

Doxastic models of games incorporating deliberation-stage beliefs were recently introduced in Bonanno (2013b, 2013c) for the analysis of dynamic games. These models allow for a definition of rational choice that is free of (subjective or objective) counterfactuals. Space limitations prevent us from going into the details of these models.

8.8 Conclusion

Deliberation requires the evaluation of alternatives different from the chosen one: in Aumann's words (1995: 15), "you must consider what other people will do if you did something different from what you actually do." Such evaluation thus requires the use of counterfactuals. With very few exceptions (discussed in Section 8.6), counterfactuals have not been used explicitly in the analysis of rational decision-making in strategic-form games. We argued that objective counterfactuals are not the relevant object to focus on, since – in order to evaluate the rationality of a choice – what matters is not what would in fact be the case but what the player believes would be the case (as illustrated in the example of the mislabeled faucets in Section 8.3). Hence one should consider subjective counterfactuals. In Sections 8.4 and 8.5 we discussed two different ways of modeling subjective counterfactuals, one based on dispositional belief revision and the other on beliefs about causal possibilities and we argued that – for the analysis of strategic-form games (and the Prisoner's Dilemma in particular) – the two approaches are essentially equivalent. We identified a restriction on beliefs (condition IND_1^{subj} of Section 8.4 and the essentially equivalent condition IND_2^{subj} of Section 8.5) which in the literature has been taken, either explicitly or implicitly, to be part of a definition of rationality. This restriction requires a player not to change her beliefs about the choices of the other players when contemplating alternative actions to the chosen one. It is a restriction that has been invoked by those who claim that "Cooperation" in the Prisoner's Dilemma cannot be a rational choice (Player 1's beliefs in the Prisoner's Dilemma expressed in COR1 [Section 8.1] violate it). What motivates this restriction is the view that to believe otherwise is to fail to recognize that the independence of the players' decisions in a strategic-form game makes it causally impossible to affect a change in the opponent's choice merely by "changing one's own choice."

Is this necessarily true? In other words, are there compelling conceptual reasons why IND_1^{subj} (or the essentially equivalent IND_2^{subj}) should be viewed as a necessary condition for rational beliefs? Some authors have claimed that the answer should be negative. Bicchieri and Greene (1999) point out a scenario (an agentive analogue of the Einstein–Podolsky–Rosen phenomenon in quantum mechanics) where causal independence is compatible with correlation and thus it would be possible for a player to coherently believe (a) that her choice is causally independent of the opponent's choice and also (b) that there is correlation between her choice and the opponent's choice, such as the correlation expressed in COR_1.

In a series of contributions, Spohn (2003, 2007, 2010, 2012) put forward a new solution concept, called "dependency equilibrium," which allows for correlation between the players' choices. An example of a dependency equilibrium is (*C*,*C*) in the Prisoner's Dilemma. Spohn stresses the fact that the notion of dependency equilibrium is consistent with the causal independence of the players' actions:

> The point then is to conceive the decision situations of the players as somehow jointly caused and as entangled in a dependency equilibrium... [B]y no means are the players assumed to believe in a causal loop between their actions; rather, they are assumed to believe in the possible entanglement as providing a common cause of their actions. (Spohn 2007: 787)

It should also be pointed out that this "common cause" justification for beliefs is generally accepted when it comes to judging a player's beliefs about the strategies of her opponents: it is a widely held opinion that it can be fully rational for, say, Player 3 to believe –in a simultaneous game – (a) that the choices of Player 1 and Player 2 are causally independent and yet (b) that "if Player 1 plays *x* then Player 2 will play *x* and if Player 1 plays *y* then Player 2 will play *y*." For example, Aumann (1987: 16) writes:

> In a game with more than two players, correlation may express the fact that what 3, say, thinks that 1 will do may depend on what he thinks 2 will do. This has no connection with any overt or even covert collusion between 1 and 2; they may be acting entirely independently. Thus it may be common knowledge that both 1 and 2 went to business school, or perhaps to the same business school; but 3 may not know what is taught there. In that case 3 would think it quite likely that they would take similar actions, without being able to guess what those actions might be.

Similarly, Brandenburger and Friedenberg (2008: 32) write that this correlation in the mind of Player 3 between the action of Player 1 and the action of Player 2 "is really just an adaptation to game theory of the usual idea of common-cause correlation."

Thus Player 1's beliefs expressed in COR_1 might perhaps be criticized for being implausible or farfetched, but are not necessarily irrational.

9 The Tragedy of the Commons as a Voting Game

Luc Bovens

9.1 Introduction

Tragedies of the Commons are often associated with n-person Prisoner's Dilemmas. This is indeed in line with Garrett Hardin's presentation of the story of the herdsmen whose cattle overgraze a commons in his seminal article "The Tragedy of the Commons" (1968) and Robyn Dawes's analysis of this story (1975). Bryan Skyrms (2001) argues that public goods problems often have the structure of n-person Assurance Games whereas Hugh Ward and Michael Taylor (1982) and Taylor (1987, 1990) argue that they often have the structure of n-person Games of Chicken. These three games can also be found as models of public goods problems in Dixit and Skeath (1999: 362–367). Elinor Ostrom (1990: 2–3) reminds us that the Tragedy of the Commons has a long history that predates Hardin starting with Aristotle's *Politics*. I will present three classical Tragedies of the Commons presented in Aristotle, Mahanarayan, who is narrating a little known sixteenth-century Indian source, and Hume. These classical authors include four types of explanations of why tragedies ensue in their stories, viz. the Expectation-of-Sufficient-Cooperation Explanation, The Too-Many-Players Explanation, the Lack-of-Trust Explanation, and the Private-Benefits Explanation. I present the Voting Game as a model for public goods problems, discuss its history, and show that these explanations as well as the stories themselves align more closely with Voting Games than with Prisoner's Dilemmas, Games of Chicken, and Assurance Games.

The support of the Economic and Social Research Council (ESRC) is gratefully acknowledged. The work was part of the programme of the ESRC Centre for Climate Change, Economics and Policy and the Grantham Research Institute on Climate Change and the Environment. I am also grateful for discussion and comments from Jason Alexander, Claus Beisbart, Veselin Karadotchev, Wlodek Rabinowicz, Rory Smead, Katherine Tennis, Jane von Rabenau, Alex Voorhoeve, Paul Weirich, and an anonymous referee.

9.2 The Tragedy of the Commons and the *n*-person Prisoner's Dilemma

Hardin (1968) argues that, just like herdsmen add cattle to a commons up to the point that is beyond its carrying capacity, the human population is expanding beyond the earth's carrying capacity. In both cases, there is an individual benefit in adding one animal or one human offspring, but the costs to the collective exceed the benefits to the individual:

> The tragedy of the commons develops in this way. Picture a pasture open to all. It is to be expected that each herdsman will try to keep as many cattle as possible on the commons. Such an arrangement may work reasonably satisfactorily for centuries because tribal wars, poaching, and disease keep the numbers of both man and beast well below the carrying capacity of the land. Finally, however, comes the day of reckoning, that is, the day when the long-desired goal of social stability becomes a reality. At this point, the inherent logic of the commons remorselessly generates tragedy.
>
> As a rational being, each herdsman seeks to maximize his gain. Explicitly or implicitly, more or less consciously, he asks, "What is the utility to me of adding one more animal to my herd?" This utility has one negative and one positive component.
>
> 1) The positive component is a function of the increment of one animal. Since the herdsman receives all the proceeds from the sale of the additional animal, the positive utility is nearly +1.
> 2) The negative component is a function of the additional overgrazing created by one more animal. Since, however, the effects of overgrazing are shared by all the herdsmen, the negative utility for any particular decision-making herdsman is only a fraction of −1.
>
> Adding together the component partial utilities, the rational herdsman concludes that the only sensible course for him to pursue is to add another animal to his herd. And another; and another... But this is the conclusion reached by each and every rational herdsman sharing a commons. Therein is the tragedy. Each man is locked into a system that compels him to increase his herd without limit – in a world that is limited. Ruin is the destination toward which all men rush, each pursuing his own best interest in a society that believes in the freedom of the commons. Freedom in a commons brings ruin to all. (Hardin 1968: 1244)

Let us construct an n-person Prisoner's Dilemma. Suppose that each player derives the same payoff from cooperating and the same payoff from defecting

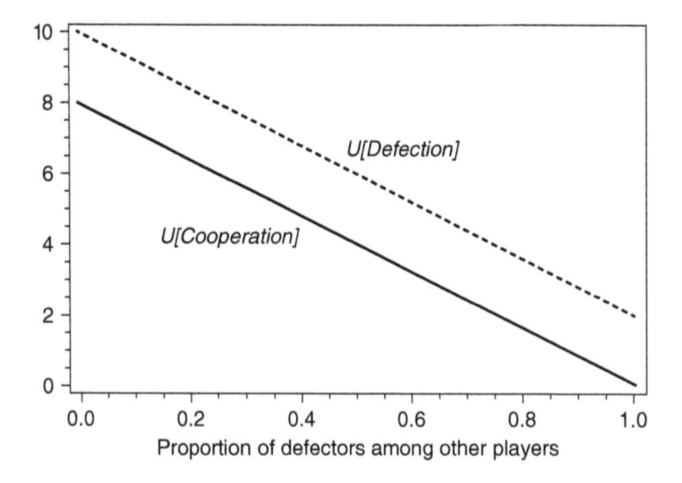

Figure 9.1 *N*-person Prisoner's Dilemma

when j other players cooperate and $(n - 1 - j)$ other players defect. In Figure 9.1, we plot the payoff from cooperating – *U[Cooperation]* – and the payoff from defecting – *U[Defection]* – as a function of the proportion of $(n - 1)$ players defecting.

No matter what strategy the other players are playing, it is better for an individually rational player to defect than to cooperate. Hence, there is a single equilibrium of All Defect in this game. But each person's payoff in the equilibrium profile is lower than each person's payoff in many other profiles, e.g. in the profile in which all cooperate, but also in some profiles with mixed cooperation and defection. Hence, the profile in which each person plays her individually rational strategy is strongly suboptimal. That is, there are profiles whose payoffs are strictly preferred by all to the payoffs in the equilibrium profile. This brings us to the core lesson of the Prisoner's Dilemma, viz. individual rationality does not lead to collective rationality. If there were a single center of control we would certainly not choose the equilibrium profile, since it is strongly suboptimal relative to some other profiles.

We could fill in the story of the herdsmen so that it is accurately represented by Figure 9.1. Defectors are herdsmen who each send an animal to the commons to feed. Cooperators refrain from doing so – e.g. they barn feed their animal. And their respective payoffs are precisely as is laid out in the graph. To do so, we need to make one assumption which is slightly unnatural, viz. that there is a negative externality of defection (commons feeding) on cooperation (barn feeding). This may be the case, but it is more

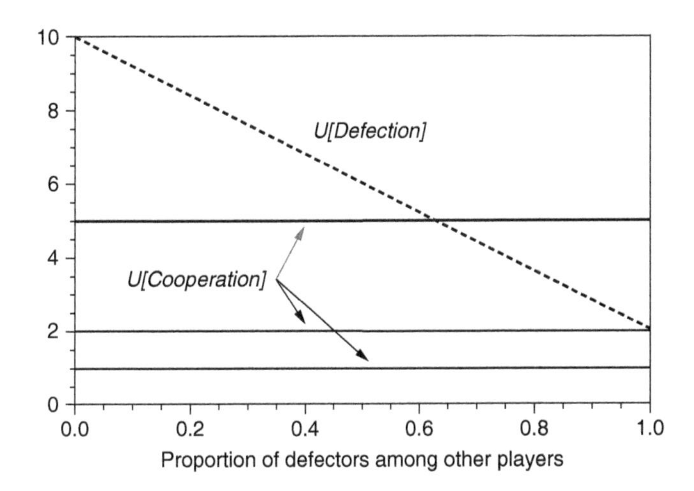

Figure 9.2 Hardin's herdsmen with fixed payoff for barn feeding

natural to think of barn feeding as having a fixed payoff. If we assume that there is a fixed payoff, then there are three possibilities, as illustrated in Figure 9.2:

(1) The individual payoffs of the All Defect equilibrium are greater than the payoff of barn feeding (bottom horizontal line), in which case the equilibrium profile is not strongly suboptimal relative to any other profile. Hence we no longer have a Prisoner's Dilemma.
(2) The individual payoffs of the All Defect equilibrium equal the payoff of barn feeding (middle horizontal line), in which case the equilibrium profile is not strongly suboptimal relative to any other profile (though it is weakly suboptimal relative to all other profiles.) This is a degenerate Prisoner's Dilemma.
(3) The individual payoffs of the All Defect profile are lower than the payoff of barn feeding (top horizontal line), in which case this profile is no longer an equilibrium. Hence we no longer have a Prisoner's Dilemma but rather a Game of Chicken, as will become clear below.

A more natural interpretation of the game in Figure 9.1 is littering. However many other people litter (Defect), each person prefers to litter rather than carry their empty beer cans back home (Cooperate). But with each additional item of litter, the town becomes less and less pleasant for litterer and non-litterer alike. This example is actually a clearer example of a Tragedy of the Commons that fits the mold of an n-person Prisoner's Dilemma.

9.3 The Assurance Game

Skyrms (2001) takes the following quote from Rousseau's *A Discourse on Inequality*:

> If it was a matter of hunting a deer, everyone well realized that he must remain faithful to his post; but if a hare happened to pass within reach of one of them, we cannot doubt that he would have gone off in pursuit of it without scruple...

Following Rousseau, an Assurance Game is also named a "Stag Hunt." In this game, it is beneficial to cooperate assuming that others cooperate as well. However, if I cannot be assured that others will cooperate then it is better to defect. That is, if I can trust that others will hunt deer with me, then I am better off hunting deer with them; but if they do not stay by their station and decide to lash out and hunt hare, then I am better off abandoning my station as well and hunt hare.

We present the n-person version of the Assurance Game in Figure 9.3. In this game, it is better to cooperate when there is a low proportion of defectors and it is better to defect when there is a high proportion of defectors. The game has two equilibria, viz. All Cooperate and All Defect. The payoff to each in the All Cooperate equilibrium is higher than in the All Defect equilibrium.

To interpret this in terms of a Tragedy of the Commons, we could stipulate that people can either choose to jointly work a commons with the promise of a large harvest if all participate or work some small garden plots by

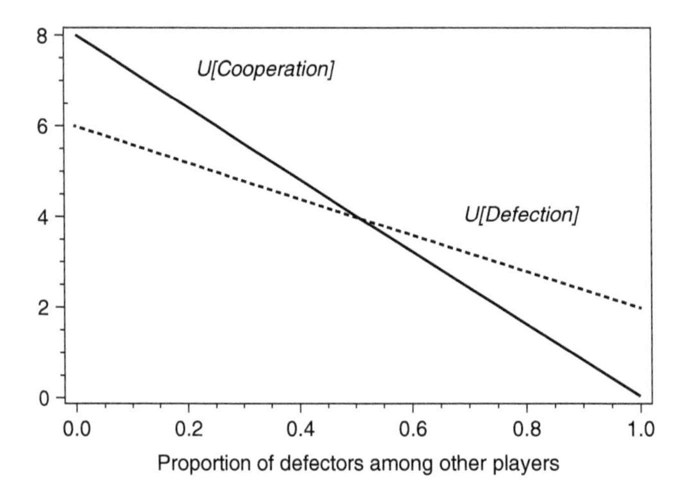

Figure 9.3 Assurance Game

themselves with the sure outcome of a small harvest. If we cannot be assured that others will cooperate then the commons will remain untapped. In this case the Tragedy of the Commons comes about due to neglect rather than over-usage, due to lack of care rather than to depletion. In reality tragedy will often ensue due to a combination of usage and lack of proper care – i.e. due to unsustainable usage.

Why might the tragedy, i.e. the All Defect equilibrium, come about in the Assurance Game? It is not because the players choose individually rational strategies as in the Prisoner's Dilemma, since it is not the case that no matter what other players do, it is better to defect. If a sufficient number of other players cooperate, then it is better to cooperate as well. The only reason why one might be hesitant choosing cooperation is because of a lack of trust – the players may not trust that others will choose to cooperate. In the n-person Assurance Game, if we cannot trust a sufficient number of people to choose cooperation, then we should just resort to defecting. And we may not trust them to do so, because, as in the Stag Hunt, we may believe that they will be blinded by short-term advantage, or there may just be a lack of trust in society at large.

9.4 The Game of Chicken

In the Game of Chicken two cars drive straight at each other forcing the other one to swerve. If one drives straight and the other swerves then the one who drives straight wins and the one who swerves loses. If they both swerve then it's a tie. And clearly, if they both drive straight then tragedy ensues. Let defection be driving straight and cooperation be swerving. Then the best response to defection is cooperation and the best response to cooperation is defection. In the n-person Game of Chicken (Figure 9.4), at low levels of defection, defection has a higher payoff than cooperation. At high levels of defection, cooperation has a higher payoff than defection.

It is easy to make this game into an interpretation of a Tragedy of the Commons. In *Fair Division and Collective Welfare*, Hervé Moulin (2004: 171) presents a case of overgrazing a commons with a fixed payoff for the cooperative strategy of barn feeding. We extend the number of players from Hardin's case. For low proportions of defectors (i.e. commons users), the same holds as in the Prisoner's Dilemma: Defecting (commons feeding) trumps cooperating (barn feeding). But for high proportions of defectors, the commons is so depleted that one would incur a loss by bringing one's animal to the commons. Hence cooperation (barn feeding) trumps defection (commons

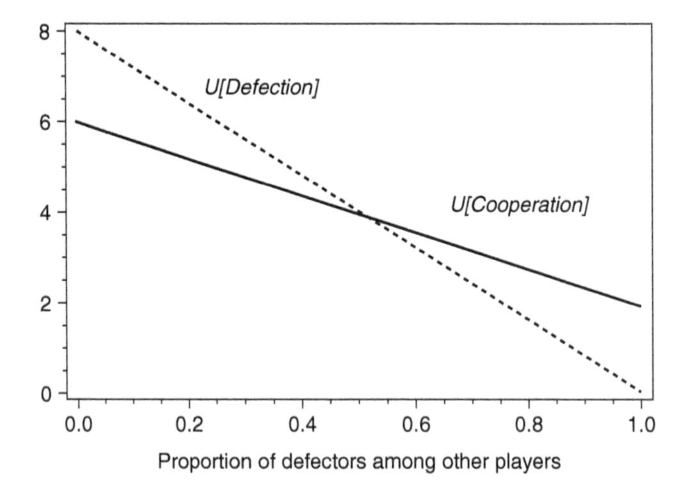

Figure 9.4 Game of Chicken with Depletion of the Commons

feeding). So by simply extending the set of players, we can turn an n-person Prisoner's Dilemma representation of an overgrazing case into a Game of Chicken representation. Case (3) in Section 9.2, represented in Figure 9.2, with a fixed barn feeding payoff higher than the payoff of All Defect, also fits this mold.

If we envision an n-person Game of Chicken with cars racing at each other from various sides, then defection (going straight) is only rational at a *very* high level of cooperation (swerving) – viz. when everyone else swerves – assuming that swerving avoids a collision. As soon as one other person goes straight, swerving is the rational response. So the intersection point between the line for defection and cooperation is between 0 and 1 defector. But this is just an artifact of the story. In the commons example, defecting (commons feeding) remains rational at high and mid-range levels of cooperation (barn feeding). We can put the intersection point between the line for defection and cooperation at any point in the graph and tell a fitting story.

In an n-person Game of Chicken, the equilibria in pure strategies are the profiles at the intersection point of the lines for cooperation and defection (at least for the continuous case). In low-level defection profiles, i.e. profiles to the left of the intersection point, it is better for a cooperator to unilaterally deviate to defection. In high-level defection profiles, i.e. profiles to the right of the intersection point, it is better for a defector to unilaterally deviate to cooperation. Only the profiles at the intersection point are equilibria.

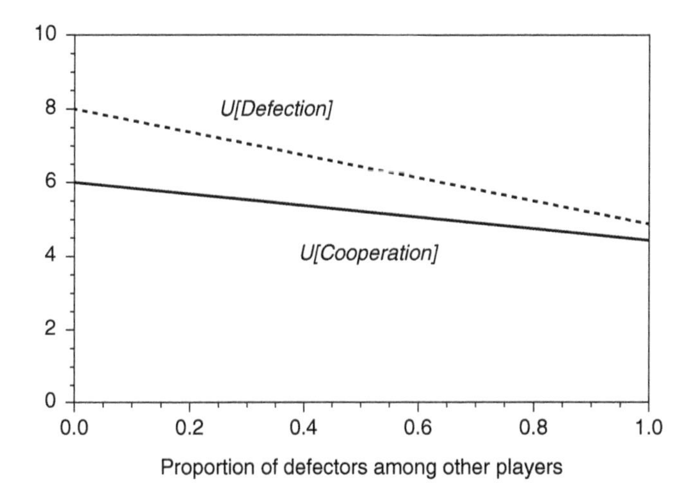

Figure 9.5 Prisoner's Dilemma as a truncation of the Game of Chicken in Figure 9.4

How does tragedy ensue? There are two reasons. First, the equilibria are not welcome equilibria. In Moulin's example of the overgrazing of the commons, the intersection point is the point at which the commons are so depleted that it makes no difference whether the next herder brings his animal to the commons or barn feeds. Let the intersection point be at $m < n$ players. If we truncate the game to m players, then we can see that the game is like the n-person Prisoner's Dilemma as in Hardin's Tragedy of the Commons. (I have represented the truncated game in Figure 9.5 with $m = 0.4n$, i.e. if there are, say, 100 herdsmen in the original game, then there are 40 herdsmen in the truncated game.) In the non-truncated Game of Chicken, if the slope of the cooperation line is negative (as it is in Figure 9.4) then the outcome of All Cooperate is strongly Pareto-superior to the outcome of any of the equilibria in pure strategies located at the intersection point.

Second, there is a coordination problem. When the herdsmen have to choose independently to bring their animal to the commons, they cannot coordinate who will and will not send their animal to the commons. They have no ground to prefer one of the equilibria in pure strategies over another. If they resort to an equilibrium in randomized strategies, i.e. they all choose to send their animal to the commons with a particular probability, then we may be lucky and end up with a number of commons feeders that is lower than the depletion point, but we may also be unlucky and end up with a number of commons feeders that is higher than the depletion point. Similarly, in the actual Game of Chicken, our luck runs out when all defect, i.e. go straight.

9.5 Three-person games

Let us consider three-person games for the three types of games that can represent Tragedies of the Commons. Start with the three-person Prisoner's Dilemma in Table 9.1. There is a persistent preference for defection over cooperation ($D \succ C$) in the Prisoner's Dilemma, whether none, one, or both of the other two people are defecting. The equilibrium is at All Defect. To construct an Assurance Game, we can turn around one preference, viz. $C \succ D$ when none of the other players are defecting. The equilibria are All Cooperate and All Defect. To construct a Game of Chicken, we can turn around one preference, viz. $C \succ D$ when all of the other players are defecting. The equilibria are Two Defect and One Cooperate.[1]

This construction of three-person games invites the following question: So what if we turn around the central preference, viz. $C \succ D$ when just one of the other players is defecting, all other things equal? This would turn the game into a Voting Game. Suppose that we have three players who would all vote in favor of a proposal if they would bother to show up and they need to muster just two votes to defeat the opposition. To cooperate is to show up for the vote and to defect is to stay at home. Then $C \succ D$ when there is exactly one other voter who shows up and $D \succ C$ otherwise – if nobody bothered showing up, then my vote would not suffice anyway, and, if two other people showed up, then my vote is not needed anymore. The equilibria in pure strategies in this game are Two Cooperate and One Defect or All Defect.

I have shown earlier that by truncating the set of players in a Game of Chicken it can be turned into a Prisoner's Dilemma. Similarly, by truncating the set of players in a Voting Game it can be turned into an Assurance Game. If we take out the third voter, then nobody can shirk if they would like to see the social good realized. Now there are only two equilibria, viz. the optimal one in which all show up to vote and the suboptimal one in which nobody shows up. This will become important in our discussion of classical Tragedies of the Commons in the next section.

[1] We can also construct an Assurance Game by turning around two preferences, viz. $C \succ D$ when none or one of the other players is defecting. This would not affect the equilibria. And we can also construct a Game of Chicken by turning around two preferences, viz. $C \succ D$ when one or both other players are defecting. Then the equilibria are two players cooperating and one defecting. In terms of the graphs in Figures 9.3 and 9.4, we are just shifting the intersection points.

Table 9.1 Variations of the three-person game

	Other players			
		CD		
	CC	DC	DD	Equilibria in Pure Strategies
Prisoner's Dilemma	D ≻ C	D ≻ C	D ≻ C	DDD
Assurance Game	C ≻ D	D ≻ C	D ≻ C	CCC, DDD
Game of Chicken	D ≻ C	D ≻ C	C ≻ D	DDC, DCD, CDD
Voting Game	D ≻ C	C ≻ D	D ≻ C	DDD, DCC, CDC, CCD

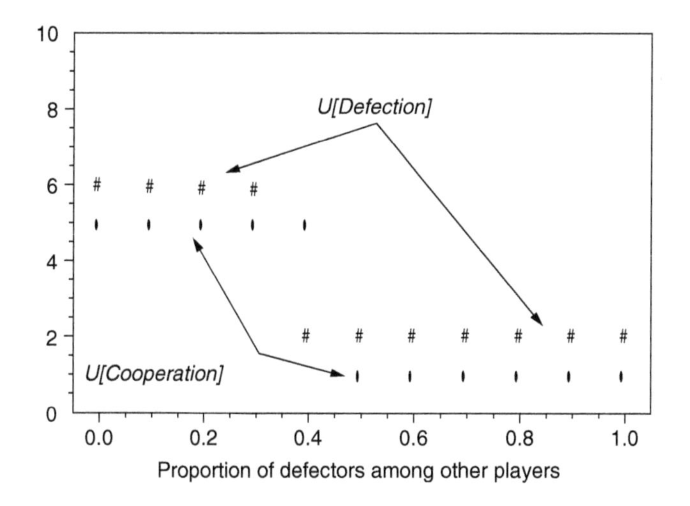

Figure 9.6 Voting Game with single pivotal player

We can construct a representation of a Voting Game in terms of the n-person game with the payoffs of defection and cooperation as a function of the proportion of other players defecting. If the game is an actual vote then there will be exactly one pivotal point at which cooperation beats defection. E.g. suppose that there are 11 voters who would all vote yes if they were to come to the polls, and we need to bring out 7 yes votes to win the vote. Then cooperation is preferred to defection by a player just in case there are precisely 6 of the 10 other voters who bothered to show up for the vote, i.e. if the proportion of defectors among the other players is 0.40. (See Figure 9.6.)

We can also construct a generalized version in which defection trumps cooperation when there are few cooperators and when there are many

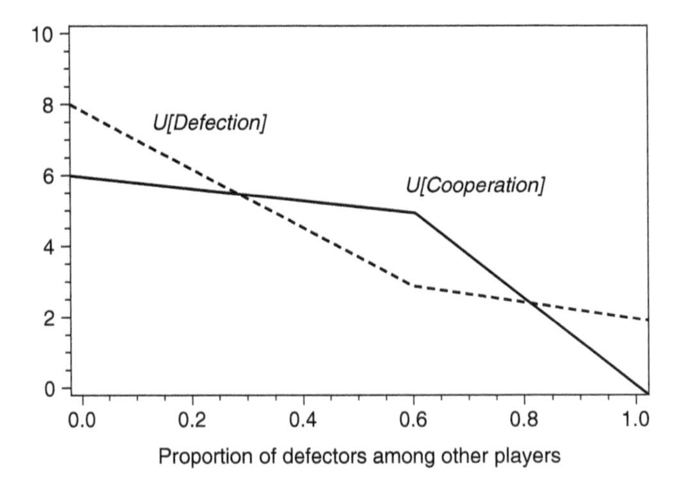

Figure 9.7 Voting Game with critical zone

cooperators, while there is a critical zone in the middle in which it is worthwhile contributing to the social good. The decreasing slopes indicate that the quality of the social good declines or that there is a decreasing chance that the social good will be realized as the defection rate goes up. The linearity assumption is just for simplicity. (See Figure 9.7.)

Voting Games do come with an air of tragedy. If the players can coordinate who will show up to vote, then all is well. They will play an equilibrium strategy with a pivotal player and the social good will be realized. But coordination is precisely the problem. When they cannot coordinate, then they may slide into the All Defect equilibrium and not show up for the vote. We also see the same pattern when voters cast a protest vote for a candidate whom they do not really want to vote in place. They trust that there will be a sufficient number of voters for an acceptable mainstream candidate. But without coordination, it may happen that there will not be a sufficient number and the protest candidate may gain unwanted traction.

There is a history of using games that have the structure of the Voting Game as an analysis of collective action problems. I will survey and discuss this literature before turning to classical Tragedies of the Commons.

9.6 Historical background of the Voting Game

The name "Voting Game" is inspired by the Simple Voting Game in Dan Felsenthal and Moshé Machover, Def. 2.1.1. in (1998: 11) which in turn builds on L.S. Shapley and Martin Shubik (1954). A Simple Voting Game is a set of voters and a non-empty set of subsets of the voters which form winning coalitions, and every superset of a winning coalition is also a winning coalition. E.g. under majority rule, the set of winning coalitions of the set of voters {a, b, c} is {{a, b}, {b, c}, {c, d}, {a, b, c}}. This Simple Voting Game induces a Voting Game as I understand it: if you belong to a minimal winning coalition, i.e. a winning coalition which is such that no proper subset of this winning coalition is itself a winning coalition, then you prefer to cast a vote rather than not to cast a vote; otherwise, you prefer not to cast a vote rather than cast a vote.

The Voting Game has the same structure as the game G3 in Taylor and Ward (1982: 356), adapted in Taylor (1987: 41), which is a model for the provision of lumpy public goods (i.e. step goods or threshold goods). Taylor and Ward (1982: 360) and Taylor (1987: 44–45) apply this game to the context of voting. Ward (1990) illustrates the game by adapting the story in Jerome K. Jerome's novel *Three Men in a Boat (To Say Nothing of the Dog)* (1889): Rowing is worthwhile if exactly one other person rows – if none row, then a single rower will make the boat circle and if the two others row, then there is no need to row. In Skyrms (2004: 117–22) the game is called "Three-in-a-Boat."

Jean Hampton (1987) also provides a game that has the same structure as the Voting Game to analyze Hume's case of Draining the Meadows. She refers to Norman Frohlich et al. (1975) and Norman Frohlich and Joe Oppenheimer (1970) as seminal work showing that collective action problems need not have the structure of n-person Prisoner's Dilemmas.

Frohlich and Oppenheimer (1970) introduce the notion of a lumpy public good and provide the by now classical example of building a bridge (1970: 109). They discuss the rationality of making donations for the provision of lumpy and non-lumpy public goods. For lumpy goods, it is rational to donate if and only if there is a chance that the donation will make a difference to the provision of the good, i.e. when the donation might be pivotal. The presentation of the games in terms of payoffs for defection and cooperation as a function of the proportion of defectors (or rather, of cooperators) goes back to Thomas Schelling (1973). Frohlich et al. (1975: 326–327) use this Schelling model (1973) to present a game in which the payoff of defection is greater at low and high level of cooperation whereas the payoff of cooperation is greater at intermediate levels of cooperation, as it is in the Voting Game. They indicate that voting is an extreme case of this game in which cooperation only pays off when there is a level of cooperation that makes a player's vote pivotal, and present it as an example of a lumpy good. The argument is revisited in Oppenheimer (2012: 60–70).

Taylor and Ward classify G3, i.e. the Voting Game, as a variant of a three-person Game of Chicken. Their argument is that in G3, as in the Game of Chicken, we have an incentive to pre-commit to defection. Now this is true in the Game of Chicken. As I indicated in footnote 1, we can construe two Games of Chicken in the three-person game, depending on the point at which there is a preference switch. If the game has the structure of the Game of Chicken in our Table 9.1, then two players have an incentive to pre-commit to defection, forcing the third to cooperate. If we turn around the preference from D $\succ$ C to C $\succ$ D in the column of other players playing DC and CD, then we still have a Game of Chicken. In this game, the first (but not the second) player still has an incentive to pre-commit to defection. The same holds, argue Taylor and Ward, in G3, i.e. in the Voting Game.

But this is not straightforward. If one person pre-commits to defection in the Voting Game, then what is left is an Assurance Game for the remaining players. And this game has an optimal equilibrium (Both Cooperate) and a suboptimal equilibrium (Both Defect). Hence, it is not guaranteed that a pre-commitment to defection will lead to the optimal equilibrium. If there is a high level of distrust, both remaining players will choose defection. Taylor will respond that rationality prescribes that players choose the optimal equilibrium in the Assurance Game (1987: 19), but it is not clear that rationality alone is sufficient to secure trust.

Hampton analyzes a three-person variant of Hume's Draining the Meadows as a game that has the same preference structure as the Voting Game, as I will do below. But she disagrees with Taylor and Ward that this game is a Game of Chicken (1987: 254 f. 7) and classifies it as a Battle of the Sexes. Her argument is that the Voting Game is a game like the Battle of Sexes in that we need to coordinate on an equilibrium – viz. we need to determine which two people are going to conduct the work. One might say that there is indeed a resemblance in this respect.

But there is a crucial point of dissimilarity. In the two-person Battle of the Sexes, there are only two equilibria and there is a coordination problem. Hampton (1986: 151) presents a three-person Battle of the Sexes to analyze the leadership selection problem in Hobbes. There are only three equilibria in this game and there is a coordination problem. Unlike in Hampton's three-person Battle of the Sexes, there is a fourth suboptimal equilibrium in the Voting Game, viz. All Defect, aside from the three equilibria which pose an equilibrium problem. Games with different equilibria should not be classified under the same name.

One may quibble about how to delineate the precise extension of different types of games in game-theory. But following the Schelling model I do not see how the Voting Game could reasonably be placed in the same category as a Game of Chicken or a Battle of the Sexes. I take it to be an essential feature of the Game of Chicken and the Battle of the Sexes in the two-person game that defection is the rational response to cooperation and that cooperation is the rational response to defection. This generalizes, for the n-person game, to Defection being the rational response to All Others Cooperate and Cooperation being the rational response to All Others Defect. This feature is missing from the Voting Game and hence it warrants a separate classification.

9.7 Three classical Tragedies of the Commons

We will now look at three classical sources that can readily be interpreted as Tragedies of the Commons. The oldest one is Aristotle's comment in the *Politics* about common ownership. Then there is a sixteenth-century Indian story relating an exchange between the Mogul Emperor Akbar and his advisor Birbar. Finally, there is Hume's famous story in the *Treatise* of the commoners who are charged with draining the meadows.

These stories long predate the advent of game theory and they can be modeled in different ways. But it is of interest to look at them closely and to assess what the authors considered to be the cause of the tragedy – i.e. what they take to be the reason why the social good is not being realized. Their diagnoses include various observations. Many of these observations are more readily interpreted in terms of a Voting Game, rather than a Prisoner's Dilemma, an Assurance Game, or a Game of Chicken.

We start with Aristotle's passage in the *Politics*. Ostrom (1990: 2) only quotes a short excerpt: "[W]hat is common to the greatest number has the least care bestowed upon it. Everyone thinks chiefly of his own, hardly at all of the common interest." What is the context of this quote? In the *Politics*, Bk II, 1261a–1262b, Aristotle discusses Socrates' suggestion that a community (of men) should own all things in common, including women and children. He discusses "a variety of difficulties" with "all of the citizens [having] their wives in common" (1261a). One of these objections is in the passage that Ostrom is quoting. Here is the complete passage:

> And there is another objection to the proposal. For that which is common to the greatest number has the least care bestowed upon it. *Everyone thinks*

> *chiefly of his own, hardly at all of the common interest*; and only when he is himself concerned as an individual. For besides other considerations, *everybody is more inclined to neglect the duty which he expects another to fulfil*; as in families *many attendants are often less useful than a few*. Each citizen will have a thousand sons who will not be his sons individually but anybody will be equally the son of anybody, and will therefore be neglected by all alike. (1261b; emphasis added)

Aristotle seems to be worried in this passage not so much about the well-being of the women, but rather of their male offspring. He also mentions that the Upper-Libyans are rumored to have their wives in common but they divide the children among the potential fathers on ground of the resemblance that the children bear to these potential fathers. (1261a)

Aristotle provides a number of reasons why cooperation is foregone under common ownership regimes in this passage.

(A.i) Common owners only attend to their private interests and not to the common interest, and,

(A.ii) They expect others to cooperate so that the social good will be realized without their cooperation.

The passage on the ideal number of attendants in families suggests the following generalization:

(A.iii) Defection is more like to occur when there are many people rather than few people expected to procure the social good.

Rohit Parikh (2009) points to a sixteenth-century Indian story from the Birbar jest-books as recorded in the nineteenth century by Mahanarayan and analyzes it as an n-person Prisoner's Dilemma. Since the passage is not well-known, I quote it in full:

> One day Akbar Badshah said something to Birbar and asked for an answer. Birbar gave the very same reply that was in the king's own mind. Hearing this, the king said, "This is just what I was thinking also." Birbar said, "Lord and Guide, this is a case of 'a hundred wise men, one opinion' ... The king said, "This proverb is indeed well-known." Then Birbar petitioned, "Refuge of the World, if you are so inclined, please test this matter." The king replied, "Very good." The moment he heard this, Birbar sent for a hundred wise men from the city. And the men came into the king's presence that night. Showing them an empty well, Birbar said, "His Majesty orders that at once every man will bring one bucket full of

> milk and pour it in this well." The moment they heard the royal order, every one reflected that where there were ninety-nine buckets of milk, how could one bucket of water be detected? Each one brought only water and poured it in. Birbar showed it to the king. The king said to them all, "What were you thinking, to disobey my order? Tell the truth, or I'll treat you harshly!" Every one of them said with folded hands, "Refuge of the World, whether you kill us or spare us, the thought came into this slave's mind that *where there were ninety-nine buckets of milk, how could one bucket of water be detected*?" Hearing this from the lips of all of them, the king said to Birbar, "What I'd heard with my ears, I've now seen before my eyes: 'a hundred wise men, one opinion'!" (Mahanarayan: 1888: 13–14; emphasis added)

The wise men say that they figured that one bucket of water could not be detected when mixed with ninety-nine buckets of milk. Hence, they explain their defection in the following way:

(M.i) They expected all others to cooperate and hence, there was no need for them to cooperate.

Finally, let us turn to Hume, who discusses the problem of draining a commonly owned meadow in the section entitled "Of the Origin of Government" in the *Treatise*, Book Three, Part II, Section VII:

> Two neighbours may agree to drain a meadow, which they possess in common; because it is easy for them to know each others mind; and each must perceive that the immediate consequence of his failing in his part, is, the abandoning of the whole project. But it is very difficult, and indeed impossible, that a thousand persons should agree in such action; it being difficult for them to concert so complicated a design, and still more difficult for them to execute it; while each seeks a pretext to free himself of the trouble and expence, and would lay the whole burden on others. Political society easily remedies both inconveniences.

We find the same theme in Hume as in Aristotle:

(H.i) It's easier to improve a commons with a few people than with many people.

His reason for this is as follows:

(H.ii) With fewer people, a single defection would be the end of the collective project.

He offers three problems when there are more rather than fewer people, viz.

(H.iii) "know[ing] each other's mind,"
(H.iv) "concert[ing] so complicated a design," and,
(H.v) "execution," that is, of preventing that each other player would "seek a pretext to free himself of the trouble and expense" and "lay the whole burden on others."

9.8 The classical tragedies as Voting Games

These three classical stories all display Tragedies of the Commons. It concerns the lack of care for commonly owned property or for a resource that, on king's orders, needs independent attention from multiple people. Each of these stories can be construed as a Voting Game. In Mahanarayan's story, when there are few buckets of milk or many buckets of milk in the well, it makes no difference to the prospect of punishment whether one pours milk or water; but there is a critical zone between too few and too many in which one bucket of milk may reasonably make a difference between incurring the wrath of the King or not. In Hume's story, one can envision a critical zone in which cooperation is worthwhile whereas defection is a rational response when there are too few cooperators to carry out the project and too many so that one can safely shirk. And a similar interpretation can be proffered for Aristotle's attendants in domestic service.

But would this be the most plausible interpretation? To assess this, we need to look carefully at the explanations of what gets us into the tragedy in these classical sources. I distinguish between four types of explanations, viz. the Expectation-of-Sufficient-Cooperation Explanation, the Too-Many-Players Explanation, the Lack-of-Trust Explanation, and the Private-Benefits Explanation.

9.8.1 The Expectation-of-Sufficient-Cooperation Explanation

Both Aristotle (A.ii) and Mahanarayan (M.i) mention that each person expects that the social good be realized through the cooperation of others and hence that their cooperation is not needed.

This is not consistent with a Prisoner's Dilemma since there is simply no reason to assume that anyone would cooperate in a Prisoner's Dilemma. It is not consistent either with a Game of Chicken: We also expect massive defection with an extended set of players in a depletion of the commons. Nor is it consistent with an Assurance Game. In an Assurance Game, we might indeed expect cooperation considering that the optimal equilibrium is salient. But if we expect cooperation in an Assurance Game, then there is precisely reason for us to cooperate and there is no reason to defect. Hence, the Expectation-of-Sufficient-Cooperation Explanation is not consistent with a Prisoner's Dilemma, a Game of Chicken, or an Assurance Game.

However, it is consistent with a Voting Game. This is precisely what people say to explain their lack of participation in democratic procedures: They expected a sufficient number of people to show up for the vote and decided that they themselves did not need to show up. In a Voting Game, there are indeed multiple equilibria that have a mix of cooperation and defection. If the players cannot coordinate their actions and make any binding agreements, then they may indeed find themselves in the predicament that too many of the players come to expect that there is enough cooperation and that their cooperation is not needed. And consequently the situation deteriorates into the equilibrium of All Defect.

9.8.2 The Too-Many-Players Explanation

Both Aristotle (A.iii) and Hume (H.i and H.ii) mention that the problem of defection is particularly acute when there are more rather than fewer players. If we read this passage in terms of bargaining theory or cooperative game-theory – i.e. how can rational players come to an agreement about which solution in the optimal set they will settle on – then it is consistent with all games. The greater the number, the harder it is to reach an agreement due to conflicting conceptions of fairness, the desire to hold out, etc.

But we should bracket bargaining theory in our interpretation of this quote. Bargaining theory enters in when Hume's "political society" enters in. The claim here is that, before political society enters in, before we sit down to settle on a collectively rational solution, the problem of defection is more acute when there are many rather than fewer individually rational persons.

So what happens when we contract the set of players in the game from many to fewer players? In a Prisoner's Dilemma, individually rational people will simply defect whether there are more or fewer players. In a Game of Chicken, contracting the set of players would do no more than take out the cooperators and we would be left with a Prisoner's Dilemma with only defectors. So neither the Prisoner's Dilemma nor the Game of Chicken can provide an interpretation of the Too-Many-Players Explanation.

But now pay close attention to (H.ii) – with fewer people, a single defection would be the end of the collective project, Hume says. It is very natural to read this passage in terms of pivotality. With fewer people, it is more likely that each person is pivotal in carrying out the social project, whereas with more people, it is more likely that some people are dispensable. So with fewer players, it is more likely that we have an Assurance Game in which each player is pivotal or in which cooperation is at least worth each player's while. We have shown how an Assurance Game can be construed as a truncated Voting Game. If we take out the surplus of players in a Voting Game for whom cooperation is no longer beneficial when enough people cooperate, then a Voting Game becomes an Assurance Game. With few neighbors to drain the meadows in Hume's example, nobody can shirk and the project will come to completion. We can read Aristotle in a similar manner: With fewer attendants in a family, there is a better chance that the domestic work will be done well.

So this is once again an argument in favor of interpreting these classical tragedies as Voting Games. When Aristotle and Hume are insisting that fewer do better than more, we can plausibly say that they see tragedy ensue when the game is a Voting Game and that the situation can be improved by truncating it into an Assurance Game with a single salient optimal equilibrium.

9.8.3 The Lack-of-Trust Explanation

Hume mentions problems of "knowing the minds of others" (H.iii) and of "execution," of others not finding "a pretext" to shirk (H.v).

If we stipulate that the players have common knowledge of rationality, then they do know the minds of others in a Prisoner's Dilemma. They know that all will defect in the Prisoner's Dilemma and so the Prisoner's Dilemma is not a good interpretation of Hume's quote. In a Game of Chicken that reflects a Tragedy of the Commons, the expectation is also that many will defect – so many that the commons will be depleted. So also a Game of Chicken is not a good interpretation of Hume's quote.

Skyrms (2001: 2) suggests an Assurance Game as an interpretation of Hume's story of draining the meadows. Now it is indeed the case that there are two equilibria, viz. All Cooperate and All Defect. So one might say that we cannot know the minds of others and be confident what equilibrium they will play. And indeed, even if we agree to cooperate, then there still is a problem of "execution," of others not finding "a pretext" to shirk (H.v). So there is indeed an issue of trust: Just as in the Stag Hunt, the farmers might ask, "How can I be confident that others will not be blinded by their short-term interests and attend to their private business rather than show up to drain the meadows?"

So admittedly an Assurance Game would be defensible in the face of the Trust Argument. The only reservation against this interpretation is that if the collective benefit is sufficiently large and the private benefit sufficiently small, then the optimal equilibrium should be sufficiently salient for the players to have at least some reason to expect that others will cooperate – they have some grounds to know the mind of others and to trust that others won't shirk.

However, knowing the mind of others and shirking are more unsurmountable problems in a Voting Game. There are multiple optimal equilibria and, not knowing the minds of others, players do not know whether they are pivotal voters. Furthermore, it is all too easy to succumb to wishful thinking and create a pretext that there are likely to be a sufficient number of cooperators. Or, it is all too easy to succumb to defeatism and say that there are bound to be too many defectors. In an Assurance Game, trust may be the problem, but at least the saliency of the optimal equilibrium would provide some reason for trust, whereas in a Voting Game the structure of the game is an open invitation for distrust.

9.8.4 The Private-Benefits Explanation

Aristotle explains the lack of care for commonly owned property on grounds of the fact that our agency is determined by our private benefits and not by the collective good (A.i).

I admit that there is a very natural way of reading this passage on the background of a Prisoner's Dilemma: Individual rationality is what leads to the suboptimal equilibrium of All Defect in the Prisoner's Dilemma, whereas if our agency were determined by collective benefits we would not get caught there. It would also be consistent with a Game of Chicken when we extend the set of players beyond the point of profitable commons use, leading to a suboptimal equilibrium of Many Defect.

However, there is also a reading of this passage that is consistent with an Assurance Game or a Voting Game. We might say that the preferences that enter into the definition of the game reflect benefits derived from the collective project and from personal projects. The benefits from the collective project are greater than from the individual project. However, we may attach more weight to the smaller benefits from the personal project than the larger benefits from the social project in our deliberation, and our agency may be determined by these differential weights. This reading could be placed in the context of an Assurance Game: The smaller benefits from hunting hare have more weight in a player's deliberations than the larger benefits of hunting deer. Similarly, we can also place this reading in the context of a Voting Game: The smaller comforts of not showing up for the vote have more weight than the larger benefit we would derive from the social project of voting the proposal in place. We could invoke the Private-Benefits argument as a kind of myopia that drives us away from the optimal equilibrium towards the suboptimal equilibrium in an Assurance Game. It also drives us away from the optimal equilibria through the lack of coordination, self-deception, and defeatism into the suboptimal equilibrium in a Voting Game. But granted, this last explanation is not decisive and is open to multiple interpretations which can be accommodated by all four games.

9.9 Conclusion

Tragedies of the Commons can plausibly be analyzed as Prisoner's Dilemmas, Assurance Games, Games of Chickens, and Voting Games. Voting Games can trace their ancestry to the voting power literature on the one hand and to the literature on the provision of lumpy public goods on the other hand. I have looked at three classical sources (Aristotle, Mahanarayan, and Hume) that cite different reasons of why tragedy ensues. These reasons can be captured in four explanations, viz. the Expectation-of-Sufficient-Cooperation Explanation, the Too-Many-Players Explanation, the Trust Explanation, and the Private-Benefits Explanation. The Voting Game is a particularly suitable candidate to analyze these classical tragedies. Not all these explanations point univocally to a Voting Game, but the balance of reasons is on the side of a Voting Game interpretation, rather than Prisoner's Dilemma, Game of Chicken, or Assurance Game interpretations.

10 The role of numbers in Prisoner's Dilemmas and public good situations

Geoffrey Brennan and Michael Brooks

> It was a special feature of Musgrave's, Olson's and . . . Buchanan's work that they stressed the theoretical importance of group size.
>
> (Tuck 2008: 3–4)

10.1 Introduction

Although the Prisoner's Dilemma was originally developed and analyzed as a two-person interaction, many of the most important applications of what we might loosely call "Prisoner's Dilemma thinking" involve issues in the social sciences that are concerned with much larger numbers.[1] This fact immediately poses a question: How does the two-person version differ from the large-number Prisoner's Dilemma? Do the lessons of (and intuitions arising from) the two-person case carry over to larger scale social applications?

The general consensus in the economics literature is that the differences are very considerable – amounting to something like a qualitative difference between small-number and large-number situations. Consider, for example, the case of market provision of so-called "public goods." As Richard Tuck observes in the epigraph, virtually all the classic writers on public goods provision make two points: first, that the public goods problem is very like the Prisoner's Dilemma problem in certain critical respects;[2] and second, that

[1] Applications from specific issues like closed shop unionism, or the appropriate treatment of intellectual property, or compulsory inoculation programs to more general issues such as voting behavior (Brennan and Lomasky [1993]) or the basic political organization of society (Buchanan [1975]; Hampton [1986]). For an application of Prisoner's Dilemma reasoning to questions of military tactics, see Brennan and Tullock (1982) and Bellany (2003), who emphasizes the numbers dimension.

[2] For example, "[t]he free-rider problem in public goods theory is an example of what may be called a 'large-number Prisoner's Dilemma', a problem that is pervasive in many areas of economic theory" Buchanan (1968/1999: 95).

small-number cases are unlike large-number cases in that voluntary action is much more likely to secure satisfactory levels of public goods provision in the small-number setting. Buchanan, for example, observes:

> the numerous corroborations of the hypothesis in everyday experience are familiar. Volunteer fire departments arise in villages, not in metropolitan centers. Crime rates increase consistently with city size. Africans behave differently in tribal culture than in urban-industrialized settings. There is honor among thieves. The Mafia has its own standards... Litter is more likely to be found on main-traveled routes than on residential streets. (Buchanan 1965/1999: 322)

There is in short consensus that numbers make a difference; but much less consensus concerning exactly how and why they do.

Our strategy in exploring this issue is to begin by stating a simple proposition that connects numbers with market failure and then demonstrating the validity of that proposition in both the Prisoner's Dilemma and public goods cases. So in Section 10.2 we state and demonstrate our proposition in both simple cases. We shall in the process try to say a little about the relevant differences between public goods and *n*-person Prisoner's Dilemma interactions. In Section 10.3, we shall examine in some detail the work of James Buchanan on this issue, since he is the author who most explicitly emphasizes the role of numbers. It will be our claim that Buchanan's treatment is based either on a logical confusion or involves appeal to considerations that lie strictly "outside" the Prisoner's Dilemma model or both. In Section 10.4, we shall explore a treatment involving explicit appeal to "norms of cooperation" and their enforcement; and try to show why the role of numbers is somewhat more ambiguous than the classic treatments have suggested.

10.2 The core proposition?

To frame our proposition, consider Paul Samuelson on the public goods case:

> One could imagine every person in the community being indoctrinated to behave like a "parametric decentralized bureaucrat" who *reveals* his preferences [for the public good] by signalling... But... by departing from his indoctrinated rules, any one person can hope to snatch some selfish benefit in a way not possible under the self-policing competitive pricing of private goods. (Samuelson 1954: 389)

Our preliminary proposition takes off from his observation. It is:

Proposition 1: The "*selfish benefit*" to which Samuelson refers is larger, the larger the relevant numbers.

If, as we believe, proposition 1 is true, then an immediate link between numbers and market failure seems to be established: Whatever forces are in play to sustain the optimal outcome, they will have to be stronger the larger the "relevant numbers."

But a question immediately arises as to *which* number is "relevant." In particular, the term might refer to the "n" in the n-person Prisoner's Dilemma, or to the number of complier/contributors, which we shall denote "c". Samuelson's original exposition tends to suppress that distinction, because it focuses on a situation where the initial benchmark output is optimal – a case where n and c are equal. But in what follows, we will be concerned to deal with more general cases, where the number of affected persons (the "n" referred to in the n-person case) and the number of contributors are different. We can illustrate the distinction by distinguishing two different questions:

1) Is market failure more likely (or alternatively more extensive) for larger number than for smaller number public goods? (This question focuses on n.)
2) If some individual I contributes to providing the public good ($c < n$), does that individual's contribution increase or decrease the incentive for others to contribute? (This question focuses on c.)

In demonstrating proposition 1, we shall aim to keep these questions apart. As indicated, we shall pursue the demonstration in three parts: first, we shall deal with a simple form of the Prisoner's Dilemma game; second, we shall extend the treatment to the more complicated public goods case; and third, we shall say a little about the differences between these two cases.

10.2.1 The n-person Prisoner's Dilemma case

Consider a standard form of the n-person Prisoner's Dilemma game as typically formulated for laboratory experiments. There are n players. Each is given an endowment of E. To simplify, each has only two possible strategies: either to contribute all her E to a common pool; or to retain it. All contributions are aggregated, increased by factor Ω (greater than 1) and then returned to all n in equal per capita amounts independent of whether the recipient contributed or not. Let the number of other contributors be c. Consider individual I. If I contributes, she receives $[\Omega(c + 1)E/n]$. If she does not contribute, she receives $[E + \Omega cE/n$ or $E(1 + \Omega c/n)]$. So, the net benefit (B) from her *not* contributing is:

$$B = E(1 + \Omega c/n) - E(\Omega c + \Omega)/n = E/n[n + \Omega c - (\Omega c + \Omega)]$$
$$B = E(1 - \Omega/n) \ldots (1)$$

It will be seen by inspection:

(1) that B is positive provided $n > \Omega$ (i.e. when the number of affected individuals exceeds the multiplier);
(2) that B is independent of c (the number of other contributors); and
(3) that B is *larger* the larger is n.

Clearly, the best outcome for all (the Pareto-optimum) is where all contribute and each receives ΩE; but each player has an incentive to depart from that outcome. In fact, there is a dominant strategy for each player, which is to contribute nothing. This zero-contribution to the common pool is the "Nash equilibrium" in this game. And so the situation exhibits the characteristic Prisoner's Dilemma property: there is a unique Nash equilibrium that is Pareto-dominated. When each contributes nothing, each receives a payoff of E; which is less than the ΩE obtainable if all cooperated.

Moreover, in this simple game, the incentive to contribute nothing is larger the larger the number of persons in the community. (The size of that incentive is independent of whether c takes its maximal value of $(n - 1)$ where all others contribute, or its minimal value where no others contribute, or anything in between.) Put another way, unless the parameter Ω, which measures the relative gain from cooperative activity, is greater than or equal to n then the payoff maximizing strategy is to not cooperate (or "defect"). So, as n rises, any forces encouraging contribution have to increase in order to make contributing rational.

10.2.2 The public goods case

The public goods analog to the Prisoner's Dilemma case is more complicated because the benefits of the public good both across all persons and to each individual are non-linear in the level of supply. But we can illustrate the Samuelson point reasonably simply in terms of a partial equilibrium diagram

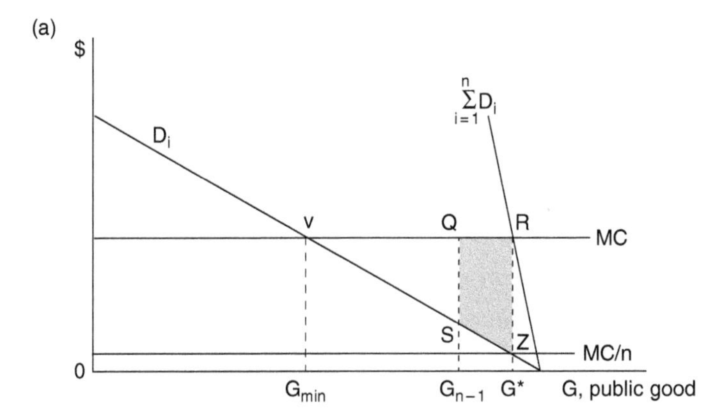

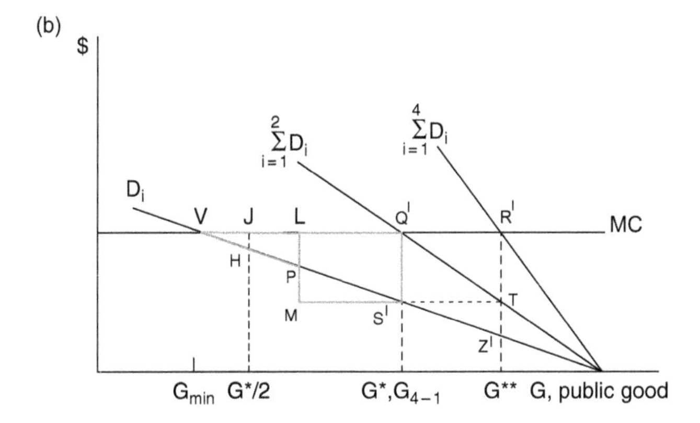

Figure 10.1 On the incentive to free ride

indicated by Figure 10.1. Suppose there are n identical agents and that the (identical) individual demand curve for the public good G is given by D_i. The public good is produced under conditions of constant average (and hence constant marginal) cost MC; and the Samuelsonian optimum is given by G^*, the quantity each would demand at "price" MC/n. We begin, in the Samuelsonian spirit, by supposing that the Pareto-optimum were somehow achieved and then determine the size of the "temptation" to defect from this starting point.

At G^*, the marginal benefit to individual I from his public goods contribution is of the order of MC/n per unit. The cost to him per unit is MC. The cost exceeds the benefit – and does so by more the larger is n. In terms of panel a of Figure 10.1, the net benefit to I of not-contributing is given by the shaded area, the excess of cost to I minus the marginal value to I of the public good his contribution makes possible.[3]

On this basis, we can clarify one misconception that Tuck (2008: 8) seems to have taken from Olson (1965). Tuck interprets Olson as claiming that, as n becomes larger and larger, the change in the public good supply is becoming smaller and smaller: that the limit of $\Delta G/n$ as n approaches infinity is zero.[4] But as the diagram makes clear, this is precisely *not* the relevant point. What tends to zero as n becomes large is the value to I of I's marginal G-*contribution*, ΔG. And this remains true however large or small ΔG is, provided that the individual's demand curve for G lies below the marginal cost curve over the relevant range (i.e. provided output lies in the range beyond G_{min} in Figure 10.1). Of course, since $G^* >> G_{min}$, then Samuelson's logic is sufficient to establish his market failure claim.

But two features of the case are worth emphasizing. First, the larger is n, the smaller is MC/n relative to MC and therefore the greater the incentive for any individual to refrain from contributing at G^* when all others are contributing. (The psychic factor encouraging contribution has to be larger,

[3] Our diagram may mislead. In a world of economic equals, as n increases the collective demand curve will become steeper and steeper approaching a vertical line, the gap in output between G^* and G_{n-1} will shrink, as will the shaded quadrilateral to virtually a line-segment drawn between MC and the marginal evaluation curve. Our diagram in which such features do not exist is best viewed as a caricature of the large-numbers case.

[4] We agree that Olson is unclear on this point and may himself have been somewhat confused. For example, Olson claims that *homo economicus* assumptions are not necessary for the free rider problem: "Even if the member of the large group were to neglect his own interests entirely, he still would not rationally contribute toward the provision of any . . . public good, since his own contribution would not be perceptible. [A] . . . rational [person], however unselfish, would not make such a futile and pointless sacrifice, but he would allocate his philanthropy in order to have a perceptible effect on someone" (Olson 1965: 64 as quoted by Tuck 2008: 8). Clearly, as n increases, the larger the total benefit a given increase in G would achieve across the n persons; so in that sense the philanthropist has *more* reason to contribute as n increases – *provided* his contribution will not lead to other individuals reducing theirs. If, on the other hand, his contribution leads others to reduce their contributions *pari passu* then the unselfish person's contribution would indeed lead to an "imperceptible" effect on total public goods supply. Whether that response by others is what Olson has in mind is totally unclear. But in the absence of some such effects, his claim seems just plain wrong (as Tuck observes). Wicksell (1896) in Musgrave and Peacock (1958: 81) makes a somewhat similar, misleading claim: "Whether [the individual] pays much or little will affect the scope of public services so slightly that for all practical purposes he himself will not notice it at all." Sorites problems might arise in public goods cases; but they are logically entirely independent of the basic market failure claim.

the larger is n).[5] And second, that for given n, as the level of G supplied rises, the larger the excess of MC over marginal evaluation. As G rises, individual marginal evaluation of G falls but MC remains constant, so the incentive to not-contribute is a positive function of the number of other contributors,[6] c. (Note that c proxies for G, because each makes a decision whether to contribute his share to providing the public good or not.) So, if individual I contributes, this *reduces* the net incentive for J to contribute – or equivalently, increases the incentive for others to defect.

We should perhaps emphasize at this point that the foregoing analysis (both of the public goods and Prisoner's Dilemma cases) depends on each agent treating the *actions* of other players as independent of his own. To be sure, I's payoff depends on what others do; but each is assumed to ignore any effects of his actions on others. In that sense, individuals do not behave "strategically": they behave "parametrically." If each player in the Samuelsonian conceptual experiment, for example, were to assume that his own failure to contribute would cause everyone else to cease contributing, then he would have "rational" reason to continue his "indoctrinated behaviour." And one thought might be that such strategic interactions are likely in small-number interactions. This is clearly an intuition that guides Buchanan's thinking (as we shall detail in the Section 10.3). But this is why it is important to note that, within the terms of the public goods structure, the effect of any one's contributing is to increase G (or c) *ceteris paribus*, and therefore (in general[7]) to *reduce* not increase the incentive for others to contribute. Prima facie, if I did take account of the effect of his actions on others, it seems at least as likely to discourage rather than encourage I to contribute.

Let us elaborate the point briefly. Samuelson does not isolate the Nash independent adjustment equilibrium in the public goods case: it is sufficient for his purposes to argue that optimality is unstable under independent

[5] If, for example, n equals two, then the net benefit from not contributing when the other is contributing is represented in panel b by area $S'HJQ'$. If n doubles, the net benefit from not contributing when all others are contributing is represented by area $S'Q'R'Z'$, which equals area $S'Q'R'T$ plus $S'TZ'$, redrawn for purpose of comparison as area $S'Q'LM$ plus LPV. The net gain from free-riding is larger when n doubles by area PMS' plus VJH.

[6] Suppose, for example, n equals four with each contributing under the Samuelsonian indoctrinated behavior one-fourth of the optimal supply. If c is two, then the net benefit from individual I's free-riding is represented in panel b by area $LQ'S'P$. If c rises to three, then individual I's net benefit from free riding increases to the amount represented by area $Q'R'Z'S'$.

[7] The best-response curve depicting the relationship between individual I's provision and the provision by others can be positively sloping if the individual's preferences for the private good are inferior. We think such a case is implausible.

adjustment. But it is easy enough to see that that Nash equilibrium will occur where each individual's marginal evaluation of G cuts MC – viz. at G_{min} in Figure 10.1. Of course, in order to achieve that equilibrium, we may have to relax the assumption that individuals have only two strategies – contribute their "fair share" of the costs at optimality, or contribute nothing.[8] Suppose we do relax that assumption – so that each contributes whatever amount she prefers – then the Nash equilibrium will occur where each of the identical individuals will contribute an identical amount $G_{min}.MC/n$.

Now, in this Nash equilibrium, each might note that there are gains to herself from behaving strategically. She might recognize that, if she failed to contribute, the level of public goods supply would lie below G_{min} (and hence that an additional unit of G would be worth more to any of the other individuals than it costs them to provide). So those others would have an incentive to increase their contributions and return G to its former level, G_{min}. In this sense, strategic behavior invites something like an "I'll-fumble-you-pay" outcome, where all wait for someone else to give in and contribute up to G_{min}.

10.2.3 Prisoner's Dilemma versus public goods interactions

Multi-person Prisoner's Dilemmas' and n-person public goods predicaments are alike in that if the optimal (universal cooperation) outcome emerged, it would not be a self-sustaining outcome for players who aim to maximize their "objective" payoffs in the game. In the standard Prisoner's Dilemma formulation, there is a dominant strategy for each player – an action that is best for each (in terms of "objective" payoffs), independent of what others do. That dominant strategy is not to cooperate/contribute. There is not necessarily a dominant strategy in the public goods case. The Nash independent adjustment equilibrium need not involve zero provision: Whether it maximizes a player's objective payoff to contribute or not will in general depend on the level of public goods supply.

Moreover, it might be argued that our formulation of the n-person Prisoner's Dilemma (in Section 10.2.1 above) makes the Prisoner's Dilemma too much like the public goods case. In the original illustration of the Prisoner's Dilemma, with confessing prisoners, there remains the possibility of differentiating across individuals. That is, in a three-person Prisoner's Dilemma,

[8] An alternative would be for each to contribute with probability G_{min}/G^* – but that would achieve the Nash equilibrium only in an expected sense.

where each can confess or not confess, *I* can turn state's evidence against no one (i.e. not confess) or against *J* (i.e. confess that *I* and *J* did it) or against *K* (i.e. confess that *I* and *K* did it) or against both *J* and *K* (confess that they all did it together). And different punishments could be allocated to the four action options. The "numbers" in the Prisoner's Dilemma can be individuated. In the public goods case, it is not possible for *I* to withdraw contributions without affecting *J* and *K* equally.

Finally, the "number that counts" in the Prisoner's Dilemma game is the number of affected persons: The larger the n-number, the larger the incentive for all players to "defect." In the public goods game, the "number that counts" is the number of compliers c – or, more accurately, the level of public goods supply.

10.3 The Buchanan approach

What our *core proposition* establishes is that any given "subjective" benefit associated with behaving "cooperatively" (in and of itself) will, in order to secure cooperative behavior, have to be larger the larger the numbers involved (in either public goods or n-person Prisoner's Dilemma settings). The temptation to defect/free ride is a positive function of the numbers involved.

But we formulated the title to Section 10.2 specifically as a question. We did this to prefigure the possibility that what we have indicated as the core proposition isn't "core" at all – that the connection between numbers and suboptimality that at least some of the commentators have in mind does not lie so much in the logic of the payoff structure as somewhere else.

This possibility is perhaps most clearly exemplified by Buchanan – though it also seems to be in play in Tuck's critique of Olson. In a variety of places and over a significant time period, Buchanan talks explicitly about the role of numbers in public goods and Prisoner's Dilemma cases. Buchanan's first statement is in a paper in *Ethics* in 1965.[9] Significant later statements appear in Buchanan (1967, 1968 chapters 2 and 5, 1978, and 1983). There are interesting variations among these various expositions – and sometimes, alternative statements that are at odds are offered within the same paper. There is, however, one clear message – numbers matter! Though *why* numbers matter has two rather different explanations in the Buchanan corpus and neither of them corresponds with our "core" proposition.

[9] Buchanan (1965: footnote 7) was aware of an early draft of Olson's book published in that same year, in which numbers play a principal role.

The first relates to the difference between strategic and parametric behavior. The thought is that in small-number interactions each individual takes into account the effects of his actions on others. Specifically, he can use his own contribution as a mechanism for inducing others to contribute. There are, as Buchanan would put it, gains from exchange on offer and in the small numbers cases the relevant "exchanges" look to be feasible – at least in principle.

The most striking example of such use is perhaps the two-person iterated Prisoner's Dilemma, where, as Axelrod famously established, players can effectively use the threat of defecting in subsequent rounds of play as a device for ensuring cooperation by the other. If both players play "Tit-for-Tat" (contingent cooperation), then each knows that the cost of defecting in round t is that the other will defect in round $(t + 1)$. Provided the game has no definite endpoint (and that the probability of continuing to a next round at any point is sufficiently high), this contingency may be sufficient to induce cooperative behaviour from both indefinitely.

Although it is far from clear what mechanism, if any, Buchanan[10] had in mind, one might interpret him as wanting to extend the two-person logic to more general small-number cases.

> In small-group situations, each potential trader is motivated to behave strategically, to bargain, in an attempt to secure for himself differentially favorable terms. At the same time, he will also seek to promote general agreement so as to secure the advantages of the acknowledged mutual gains. (Buchanan 1968/1999: 81)

By contrast, in the large-number case:

> The individual... will not expect to influence the behavior of other individuals through his own actions. He will not behave strategically... He accepts the totality of others' action as a parameter for his own decisions, as a part of his environment, a part of nature, so to speak, and he does not consider this subject to variation as a result of his own behavior, directly or indirectly. (Buchanan 1968/1999: 81–82)

Now, it is one thing for individuals in small-number cases to behave strategically and another entirely for those strategic interactions to produce

[10] See Buchanan (1965/1999 and 1968/1999: chapter 5), in which he basically asserts that in small-number situations an individual's contribution can increase the likelihood others will also increase their contribution. Buchanan claims that in this respect "small-numbers are beautiful" but the mechanism involved is simply not explained.

cooperative behavior. In the rather special case of two-person interactions, Tit-for-Tat may work to produce universal cooperation. Moreover, there is an important independent reason for thinking that two-person Prisoner's Dilemma's are not especially problematic. As the formulation in Buchanan (1978) explicitly recognizes, two-person exchange in private goods can be formulated as a two-person Prisoner's Dilemma: both involve a mix of conflict (as each seeks the best deal for himself) and mutual benefits. Each, by failing to deliver on his side of the bargain, can achieve a one-round advantage to himself, just as if "... they are villagers by a swamp, to use David Hume's familiar example" (Buchanan 1978/1999: 365). If players in two-person Prisoner's Dilemmas routinely fail to cooperate, that would be bad news for the operation of markets. Equally, however, it is clear that market exchanges do work tolerably well, even in situations where legal enforcement is minimal.

In any event, as Buchanan emphasizes, the contrast between the private exchange case and the public goods case relates to what happens as numbers expand. Increased numbers in the private goods cases serve to make the terms of trade independent of any individual action: large numbers eliminate the bargaining problem. But in the public goods/Prisoner's Dilemma cases, larger numbers don't solve the problem: in these cases, there is simply no alternative to "explicit genuine *n*-person agreement."

The argumentative strategy Buchanan adopts in these various expositions is to show that, since each can be taken to behave parametrically in the large-number case, it is only in small-number cases that optimality can be secured. This is a somewhat unsatisfactory argument because nothing Buchanan says suggests why in the small-number cases cooperation is especially likely to emerge. Small numbers (strategic behaviour) may be *necessary* for cooperation – but nothing in the logic of the interaction itself provides any reason for thinking that small numbers will be *sufficient*. The question is whether there is anything within the standard interaction that could induce any player to think that others' contributions are likely to be increased if she herself contributes. In the Prisoner's Dilemma game illustrated in Section 10.2, there is no such implication: The incentive to contribute is independent of my level of contribution. And in most public goods cases, the incentive to contribute is *negatively* related to the contributions of others.

Although it is far from clear what Buchanan had in mind, his argument is most plausibly interpreted as making appeal to another, apparently independent, consideration: the relation between numbers and certain independent psychological features of small number interactions. Suppose, for example, we introduce an additional payoff S as an independent "subjective" return associated with playing the cooperative strategy for its own sake. Then we could state our earlier results in terms of the claim that as numbers increase S must increase if public goods/Prisoner's Dilemma problems are to be solved. But Buchanan suggests that there are good reasons for thinking that S will diminish as numbers increase. As he puts it:

> there are... *internal* limits or bounds on what we might call an individual's moral-ethical community. There are, of course, no... categorical lines to be drawn between those other persons whom someone considers to be "members of the tribe" and those whom he treats as "outsiders"... I assert only that, for any given situation, there is a difference in an individual's behavior toward members and non-members, and that the membership lists are drawn up in his own psyche. (Buchanan 1978/1999: 364)

More to the point, perhaps, the "tribe" in Buchanan's conception is limited in numbers.

> An increase in population alone reduces the constraining influences of moral rules. Moreover population increase has been accompanied by increasing mobility over space, by the replacement of local by national markets, by the urbanization of society, by the shift of power from state-local to national government, and by the increased politicization of society generally. Add to this the observed erosion of the family, the church, and the law – all of which were stabilizing influences that tended to reinforce moral precepts ... (Buchanan 1978/1999: 365–366)

Now, we do not doubt that such social psychological factors are relevant in determining the extent to which individuals behave according to moral/social norms. But we are suspicious about wheeling them in to make an argument for large-number/small-number differences in PG (public goods) cooperation, at least if one looks to Prisoner's Dilemma logic to help *explain* why cooperative impulses might be stronger in smaller groups, as Buchanan seems to do. Indeed, Buchanan's argument looks as if it goes the following way: People behave more cooperatively in small-number than in large-number Prisoner's Dilemma settings because they behave more cooperatively in small-number than in large-number Prisoner's Dilemma settings[11] – which hardly qualifies as an explanation of anything.

[11] In Buchanan (1965/1999: 313) the ambition seems to be to deploy Prisoner's Dilemma reasoning to the choice between "alternative ethical rules": on the one hand, the "moral

It might be useful at this point to summarize what we see as the issues arising from the Buchanan treatment of numbers.

First, "making an appreciable difference to others' payoffs" (e.g. level of public goods supply) is seen to be a *necessary* condition for achieving optimality in Prisoner's Dilemma or PG situations. The point in both cases is that necessity tells us nothing in and of itself. We need reasons to think that, if the necessary conditions were met, optimality would indeed be forthcoming. But according to the internal logic of the game, as it stands – the idea of Nash equilibrium behavior in all cases – "strategic" considerations have the effect that contributions are negatively related. Although parametric behavior won't produce optimality, there is no reason to think that strategic behaviour in the Prisoner's Dilemma/PG games will either. In the strategic case, "reaction curves" are in general negatively sloped – as Buchanan himself argued in a variety of places.[12]

Second, in order to explain cooperation in the small numbers case we need to invoke considerations that lie "outside the Prisoner's Dilemma model"[13] – and in game theory terms that simply means that the representation as a Prisoner's Dilemma/PG game is incomplete. One possibility seems to be to appeal to psychological features of small (as distinct from large) number

law"; on the other, a "private maxim." Buchanan attempts to make it clear that the choice "should not be interpreted as one between 'moral' and 'immoral' rules", and we might suppose that the "private maxim" includes some ethical element. In later versions, Buchanan (1978), for example, the ambition changes to one of explaining the choice between moral versus expedient action.

[12] Buchanan does not derive the slope of the reaction curve – the slope is implied from the specific preferences featured in his standard case represented in his Figure 1 (Buchanan 1967/2001: 230 and 1968/1999: 13).

[13] Buchanan says (1965/1999: 321) of the Prisoner's Dilemma that the failure to "cooperate" arises "because of the absence of communication between the prisoners and because of their mutual distrust." But we take it most commentators would think this a mischaracterization of the problem. It is not the other's untrustworthiness that leads him to defect: it is the other's rationality when faced with a strictly dominating strategy. And communication between prisoners is neither here nor there because it cannot change the Nash equilibrium without some change in the basic payoff structure. (Empirically, communication does seem to make a difference; but this must be because communication injects into the relation some additional psychic payoffs – something which cannot be explained in terms of the rationality of players but instead by some other psychological mechanism that the objective payoffs in the game do not reflect.) Buchanan goes on to claim: "The large-number dilemma is a more serious one because no additional communication or repetition of choices can effectively modify the results" (321). But that claim too seems odd. It seems that adding additional numbers does not increase the likelihood of optimality. Whether that results in a "more serious" case of market failure is an additional claim that has not been established.

interactions – but to the extent that one takes those psychological factors to be number-sensitive it is *that* fact that is doing the explanatory work. And within the Buchanan scheme that fact is simply asserted, not derived or argued for (beyond a few casual empirical gestures).

Third, once one has moved "outside the Prisoner's Dilemma/PG model" in this way, the idea that "having a noticeable effect" on PG supply (or alternatively on others' payoffs within the Prisoner's Dilemma) is necessary for an agent's actions to be noticeable more broadly seems to involve a confusion between "within model" and "outside model" considerations.[14] A simple example illustrates. Suppose there is a widespread norm to the effect that you should wash hands after going to the toilet: Failing to do so is "unclean." Now, whether you do wash your hands after going to the toilet will be detectable in certain contexts (specifically when you go to a public toilet and there are other users present). Under those conditions, your compliance or otherwise with the prevailing norms is noticeable. But it is not noticeable because of the detectable impact that your washing (or failure to wash) has on the aggregate level of germ-spreading. The fact that your particular action is of very small significance in the promotion of disease is irrelevant: Observers nevertheless regard you as a filthy person for failing to observe the norms of hygiene – and the disesteem associated with attitude may well be sufficient to induce you to comply with the norm.[15] Moreover, the idea that people wash their hands because they think this will make other people wash *their* hands (and thereby have an appreciable effect on aggregate disease risk) seems to get the motivation quite wrong. The aggregate purpose that is served by the norm may not be totally irrelevant in explaining the justification of the norm

[14] Our terminology of subjective/objective and inside/outside the model raises some interesting conceptual issues. We do not deny that all payoffs are subjectively experienced. Everything that is relevant for behavior has to be subjective. And if game theory is going to be used to explain behavior, then there can't be any "inside/outside" the model. That is, once we have some "outside the model" motivation we have to put it "inside the model" to explain behavior. On the other hand, if we are to use game theory to illuminate particular situations (e.g. littering of beaches, or compulsory union fees, or tax evasion, or military tactics) then we as observers have to make some assumptions about the structures of the game's payoffs and agent motivations that establish the relevance of the game's structure. One way to think of the distinctions is to use "subjective" payoffs when explaining behavior and "objective" payoffs to justify that the game is ostensibly a PD game or some other game. Such a step only stakes out some of the relevant distinctions but we do not have the space to lay bare all the conceptual issues.

[15] There is evidence it does have this effect: Munger and Harris (1989) show that the proportion of toilet users who hand-wash roughly doubles if someone else is present in the public facility.

as a norm, but it does not seem to be an especially important motive in inducing compliance. And we can see this by noting the number of norms that have remained in force long after they have ceased to perform the social function that might have originally justified (and/or may have initiated) them.[16]

To the extent that the connection between numbers and "cooperation" in Prisoner's Dilemma/public goods settings involves an appeal to the role of numbers in determining the strength of relevant social/moral norms, we had better have an analytical apparatus that can say something about that relation. This is a matter we turn to in the ensuing section. However, we should emphasize that nothing we shall say in that section alters our "core proposition." It remains the case that, whatever the norms-based incentive to cooperate and however it reacts to numbers, the norm-related incentive has to overcome a rival incentive to defect which *does* depend on numbers. The core proposition remains as one piece of a more extended analysis in which social and moral norms play their proper part.

10.4 Norms and numbers

Once we admit social or moral norms into the analysis, there seem to be three questions we might pose:

1. Does the number of affected persons influence the strength of incentives to follow the relevant norms? (This is a question about the role of n, as previously defined.)
2. Does my incentive to comply with the norm increase with the number of persons who also comply (a question about c)?
3. As a variant of question 2, does my compliance with the norm increase the likelihood of others complying with the norm?

Consider 1 first. As already foreshadowed in our discussion of the hand-washing example, we draw a sharp distinction between an individual having an appreciable effect on public goods provision (or the aggregate level of cooperation) and the individual's actions being noticeable more generally. We think the main effect of increased n is in terms of the norm's *justification*.

[16] Think, for example, of the practice of shaking hands as a mechanism for incapacitating mutual sword-drawing. A more extensive catalog is provided in Brennan et al. (2013), chapter 8, where the object is to explain norms that have not so much lost their original purpose as have become anti-social.

If the benefits of your compliance with a norm are restricted to your immediate community, then the moral case for your compliance is weaker *ceteris paribus* than if everyone in the world is equally affected. Increasing the number of affected persons *n*, holding everything else equal, raises the moral stakes in complying. But as we have also argued, we do not think that those moral stakes are necessarily predominant in compliance decisions: Norms do not work mainly via their intrinsic moral appeal.[17] Admittedly, increased *n* may play some minor role as a force for increasing compliance – though we think the effects are second order: A Mother Theresa-type would be esteemed not much less if she assisted just half the number of children – the esteem she gains hinges on the exceptional nature of her performance. A dog owner who fails to clean the soiled pavement would feel not much more disesteem if his dog defecated in a crowded city center than a rural village – the disesteem accrues primarily from the observed fact. In short, the "*moral*" effects of larger numbers of affected persons are unlikely to be sufficient to overcome the rival effects captured in the "core proposition."

We therefore want to focus on questions 2 and 3 above. We begin with 3, because we have less to say about it. We should at the outset emphasize that we are dealing here with an environment in which there *are* widely held social or moral norms.[18] This means that individuals are to some extent receptive to appeals to moral considerations and/or to "the things that are done around here!" In that setting, does my compliance increase the likelihood of compliance by others?

Two considerations strike us as relevant. The first involves "exemplary behaviour." We take it that the rationale for behaving in an exemplary fashion is that others are more likely to follow the example if it is present than if it is absent. This case does not serve to explain why those others might be disposed to follow the example. It does not explain the underlying impulse itself. Rather, the exemplary behavior operates by making more salient what the implicitly shared morality requires. Exemplary action makes more vivid to others the claims of the shared normative structure. The effects may in many cases be small, but they may be significant if the exemplar is independently conspicuous (a rock star or sporting hero, perhaps).

[17] We use the term "exclusively" advisedly: We do not want to assert that they play no role at all.

[18] The distinctions between moral and social norms are an object of detailed attention in Brennan et al. (2013). Those distinctions are not relevant here: What is important rather is the idea that both exercise normative force. Both appeal to psychological considerations that in most persons are supposed to induce compliance.

The second consideration involves the force of moral admonition. If one is to encourage others to perform some act, or upbraid others for not performing it, the force of one's rebuke/encouragement is greater if the preacher is herself practising – as any parent will confirm! Here, acting in accordance with prevailing moral rules provides a form of entitlement – to speak with some authority in favor of the moral practice. Obversely, failing to act in the manner in which one encourages others to act operates as a disqualifier (or at least seriously diminishes the force of any rebuke). Of course, it is only if *I* does admonish that *I*'s compliance can affect others' behavior in this way. And again, the relevance of *I*'s admonitions and encouragements are likely to depend on how conspicuous individual *I* is. But it is worth emphasizing that the "conspicuousness" in question is not a matter of how much difference to the level of public goods output *I*'s compliance makes!

Of course, it might be claimed (in a Buchanan spirit, perhaps) that conspicuousness is a diminishing function of numbers, but there seem to be lots of cases in which this is simply not so. Sometimes a prophet's honor is more extensive outside his own country than within. One's conspicuousness as a philosopher may be negligible within one's local neighborhood but very considerable across the world of philosophers. Put another way, the amount of overlap between the groups in which one enjoys "recognition" and the groups relevant for public goods benefits may be relatively small; but as the size of the latter group increases, it tends to incorporate more of the former.

In our view, the main force influencing compliance relates to the influence of social esteem. This is a large topic which one of us has examined in some detail elsewhere.[19] Suffice it to make several claims:

1. Individuals desire to be well thought of (and avoid being ill-thought of) by others for the sake of the good opinion *in itself*.[20] And social approval/disapproval of observed actions/persons is meted out according to whether the observed agent is seen to comply with norms (among other things).
2. A crucial parameter in the effects of social esteem is the size and the quality of the audience. There is, of course, (as the hand-washing case suggests) a threshold value of audience size at zero. But more generally, as audience

[19] See, in particular, Brennan and Pettit (2004). The particular application to public goods provision is explored in Brennan and Brooks (2007a and 2007b).

[20] A good reputation may be advantageous in more material ways; but there is plenty of evidence to suggest that individuals care about the opinion of others even where they are operating in environments where no reputational effects can ensue (as in the hand-washing-case).

size increases – as, for example, one moves from anonymous cases, to one where reputation comes into play – the esteem/disesteem incentives are magnified.

3. The amount of esteem one derives from norm compliance is likely to be a function of how many others also comply. In some cases, where compliance is low[21] – cases where compliance is saintly or heroic – more compliance may *reduce* marginal social esteem: Compliance becomes less saintly or heroic the more extensively it is practiced. But in normal cases of reasonably high compliance, enforcement comes more by avoidance of disesteem than by garnering positive esteem. This seems to be the case in the hand-washing example: a person becomes noticeable and hence a target for disapproval by *failing* to wash her hands. And the level of disapproval becomes more intense the less common the failure to wash one's hands is (or, more accurately, is believed by the observer to be). In such high-compliance cases, *I*'s compliance makes hand-washing yet more common and thereby increases the disesteem-related incentive for others to comply. Again, we do not reckon that securing this effect on others' likely behavior is usually a significant consideration in *I*'s decision whether or not to comply: The overwhelmingly predominant motive is the esteem/disesteem *I* directly accrues. Nevertheless, as compliance levels increase, so does the esteem-incentive and hence compliance levels. The resultant behavioral equilibrium will not normally result in universal compliance because there will predictably be some individuals for whom compliance itself is very costly, or who place a very low value on the esteem/disesteem of others. But over a substantial range, incentive to comply is positively related to the aggregate compliance level.
4. There is a countervailing aspect that deserves mention. We have already mentioned the difference that identifiable and independent reference can make to the extent of effective audience. An individual who is well-recognized in his local village may be virtually anonymous in New York

[21] One may think that compliance cannot be low if the practice is to be a genuine "norm" – that by definition a norm is something that is standardly complied with. But consider a case where pretty well everybody believes that a practice *X* is highly commendable and believes this partly because that practice is very difficult – and hence rarely performed. There can be considerable social esteem on offer precisely for that reason. Heroic conduct in military engagement might be one example. Someone who complies with the highest standards of courage is the object of considerable esteem (and esteem-related rewards like military decorations). That person has, as we might put it, complied with a norm "more honored in the breach" – but the fact of being rarely honored does not in our lexicon disqualify the practice as a norm.

City. The hand-washing experiment already referred to was, as a matter of fact, conducted in the New York City public toilet system. The results depend on a distinction between observed and unobserved (but otherwise presumed unidentifiable) agents. The conjecture might be that such an experiment conducted in a local village would induce yet higher compliance among observed subjects, because non-compliers can be *recognized*, subjected to gossip, and their reputations sullied in a manner not possible in the "big city" context. In this way, larger numbers can reduce the force of esteem by rendering relations more "anonymous" in this sense. This is a familiar point – often arrayed in relation to the alleged demoralizing effects of the move from country to city during the industrial revolution. (However, there is no presumption that *unobserved* users in the local village would be any more likely to wash their hands.)

One upshot of all this is that, in most normal cases, compliance by one person will increase the incentive of other individuals to comply. In small-number cases, it may seem to the external observer that compliance is "reciprocal" – that all compliers comply *as if* they were involved in some explicitly cooperative venture. But no such reciprocity is at stake: Each agent is behaving "parametrically" in response to the aggregate situation as she finds it. The structure of interdependence involved in the "economy of esteem" is such that the "marginal disesteem benefit" of compliance is over a substantial range likely to be a positive function of the number of compliers.

To summarize, the "numbers" that are primarily relevant to norms are c, the number of other compliers, and A, the number of effective observers.[22] We do not think the number of affected persons, n, is particularly significant to compliance – whatever its significance from the standpoint of a consequentialist ethics. The issue of anonymity is clearly a complication that bears on the role of c – but one must be careful to specify how and why anonymity is relevant. There is, for example, no implication that the esteem associated with, say, enrolling for the local fire brigade would be higher in a village than in a city. Indeed, if anonymity is a problem in the city, one might conjecture that individuals would be more likely to seek opportunities to "distinguish themselves" in the city than in the village – more eager to "make a name" for

[22] This may include "indirect" observers – people who hear of your accomplishments second-hand. Lots of people, for example, esteem Einstein as a genius, even though they have not read anything of Einstein's or ever met him: it is just common knowledge that Einstein is amazingly clever and intellectual cleverness is a widely approved characteristic.

themselves in settings where their names are not automatically a matter of common knowledge.

Finally, we should emphasize that relevant norms do not have to be norms of cooperation as such. No doubt being generally cooperative is a source of positive esteem, and being "uncooperative" of disesteem. But the kinds of norms we have in mind are "cooperative" only in effect. In most cases, the norms will attach to a particular practice – hand-washing, or academic performance, or littering, or keeping promises, for example. The characteristic feature of such norms is that they apply to activities that are *simultaneously* widely approved and collectively beneficial – but they need not be approved *because* they are collectively beneficial.

There is a related normative issue. Here, we have been focusing on norms as possible solutions to free rider problems – so it is the collectively beneficial effects of norm-compliant behavior that has been our focus. But norms (and for that matter the esteem that may play a role in encouraging compliance) have a life of their own. We do not commit to the view that the moral force of norms is exhausted by collective benefits. And the esteem or disesteem that norm compliance may give rise to may well be counted as an independent source of "utility," over and above that derived from the increased public goods supply that greater norm compliance may deliver. We do not need to take a stance here on that issue.[23] Our central point here is just that once norms are brought explicitly into the analysis, larger numbers can make for less free-riding, not more. The mistake (as we see it) that public goods writers have made is to see the source of "individual conspicuousness" to lie exclusively in the difference the individual's contribution makes to the level of public goods supply. The contribution individual *I* makes to total public goods supply might be very tiny; but the fact that individual *I* has made it can be perfectly visible to observers. And it is the latter fact, as much as the former, that is relevant in determining the extent of free-riding.

10.5 Bottom line

The characteristic feature of Prisoner's Dilemma and public goods predicaments is that there is an equilibrium outcome that is Pareto-dominated by some alternative non-equilibrium outcome. Within the structure of the interaction, the equilibrium is a genuine equilibrium; and within the rules of game

[23] Though for a more detailed treatment see Brennan and Brooks (2007a and 2007b).

theory, any "solution" to the predicament depends on inserting something into the game structure that was not previously present.

In the face of this requirement, our picture of the operation of Prisoner's Dilemma and public goods predicaments involves two elements: One element is "internal" to the predicaments themselves and is exhausted by the objective payoffs that the situations themselves embody; the other is "external" to the game and involves a "subjective" payoff that lies alongside the "objective" and helps to determine individuals' behavior. We think of this latter element essentially in terms of substantive "norms" – not in general norms of "cooperation" as such but norms that govern the particular activities that Prisoner's Dilemma and public goods situations specify. This approach offers a way of making sense both of Samuelson's counterfactual "optimality" experiment and of Buchanan's more extended treatment of the role of numbers in such predicaments.

What this approach allows is that there *are* numbers effects involved in both elements. In Table 10.1 we gather and summarize the various claims.

The internal effects we attempt to capture in terms of our "core proposition." One aspect of our treatment is a clear distinction between two numbers: the number of persons affected (the n on the basis which n-person Prisoner's Dilemma's are designated and optimality conditions in the public goods case derived); and the number of compliers (c) (or more generally the level of compliance, if we allow, as we ought, for partial compliance in all cases).

Table 10.1 Incentive effects "inside" and "outside"

	Inside the model		Outside the model/ norm-based effects
	Prisoner's Dilemma case	PG case	
As n increases	the incentive to free ride increases	the incentive to free ride increases, given c	A second-order effect to most compliance decisions
As *c* increases	the incentive to free ride is independent of c	the incentive to free ride increases given n	In most cases, increased incentive for I to comply
As *A*, the number of effective observers, increases	–	–	Increased incentive to comply
Overall Effect	It all depends on the case		

Alongside the numbers effects captured in our "core proposition," there are also the numbers effects associated with the norm compliance aspect. Here, we have to allow for a third set of numbers – that which captures the size of the audience (A) observing the compliance (or otherwise) of the actor in question. The direction of various numbers effects in this external element run somewhat counter to those in the core proposition. That is, whereas in the "internal" aspect, increases in n and c lead to diminished incentives to cooperate, in the norm-related aspect, increased n and c lead to *increased* incentives to "cooperate" as do increases in A. We have noted that the A variable is complicated by reputational factors: individuals can become lost/diminished in the crowd. This latter effect may dominate in the numbers calculus (as Buchanan perhaps thought). But the possible anonymity effect is only one factor in a more complex picture. Overall, we think of the relation between compliance and numbers as a complicated balance of "external" and "internal" considerations. The standard presumption that large-number cases are universally and qualitatively more problematic than small-number cases seems to us to be at best arguable and almost certainly unjustified in at least some cases.

11 The inner struggle: why you should cooperate with people you will never meet again

Martin Peterson

11.1 Introduction

In Book IV of the *Republic,* Plato argues that the soul consists of three separate entities: reason, spirit, and appetite. In a key passage, Socrates and Glaucon discuss the inner struggle of a thirsty man who desires to drink at the same time as he feels an urge not to do so:

> "The soul of the thirsty then, in so far as it thirsts, wishes nothing else than to drink, and yearns for this and its impulse is towards this."
> "Obviously."
> "Then if anything draws it back when thirsty it must be something different in it from that which thirsts and drives it like a beast to drink. For it cannot be, we say, that the same thing with the same part of itself at the same time acts in opposite ways about the same thing."
> "We must admit that it does not."
> . . .
> "Is it not that there is something in the soul that bids them drink and a something that forbids, a different something that masters that which bids?"
> "I think so." [1]

Plato was not a game theorist. And his theory of the soul is, of course, open to different interpretations. Despite this, game theorists have good reason to take seriously the idea, which is at least a *possible* interpretation of what Plato sought to say, that agents are complex entities made up of several distinct subagents. In this interpretation, the man described by Plato faces an inner struggle between a thirsty and a non-thirsty subagent, who have opposing preference orderings. The decision to drink is an outcome of the game played by these two subagents. What the external agent decides to do

[1] 439A–C.

will depend on the structure of the game defined by the preference orderings of his internal subagents.

The aim of this chapter is not to defend any view about what Plato himself thought about the structure of the soul. My modest goal is to show that *if* we accept what I shall call the Platonic Theory of Subagents (and I shall call it "Platonic" irrespective of whether Plato actually endorsed it or not), *then* this has important implications for how we should analyze the Prisoner's Dilemma. In essence, my claim is that given that we accept the Platonic Theory of Subagents, then there is a class of single-shot Prisoner's Dilemmas in which it is rational to cooperate for roughly the same reason as it is rational to cooperate in indefinitely repeated versions of the game. This casts doubt on the orthodox analysis of single-shot Prisoner's Dilemmas, according to which it is never rational to cooperate in such games.

The key idea of my argument is that even if you and I will never meet again in future rounds of the game, my subagents play *another* Prisoner's Dilemma that determines what it is rational for me to do in the external game. My subagents play an indefinitely repeated game, and because each subagent can be punished by other subagents in future rounds if "he" does not cooperate, it becomes important for each subagent to adjust the strategy in the current round to the threat of future punishment. I will show that under realistic assumptions, each player's subagents in a two-person internal game benefit from cooperating with the other subagent, meaning that it is rational to cooperate in many (but not all) external single-shot Prisoner's Dilemmas. A central premise of this argument is that the internal games played by subagents do not change the structure of the external game played by you and me. For my argument to go through, I must be able to show that the external game is a genuine single-shot Prisoner's Dilemma rather than a game in which interactions between subagents lead to shifts in the preference orderings of some external agents.

The idea that we sometimes face internal Prisoner's Dilemmas was first proposed in a modern game-theoretical context by Gregory Kavka.[2] Kavka limits the discussion of what I call the Platonic Theory of Subagents to internal single-shot Prisoner's Dilemmas. He therefore overlooks a key feature

[2] See Sections 1 and 2 in Kavka (1991) and Kavka (1993). Somewhat surprisingly, Kavka does not mention the connection with Plato's tripartite theory of the soul. For a discussion of the psychological aspects of Kavka's proposal, see Frank (1996), who also mentions the connection with Plato's tripartite theory of the soul.

of the argument explored here – namely, the claim that internal and *indefinitely repeated* Prisoner's Dilemmas played by two or more subagents can be part of an external single-shot Prisoner's Dilemma played by you and me, in a sense that makes it rational for us to cooperate. It is this feature of my argument that is novel.

It should be noted that advocates of the Platonic Theory of Subagents need not believe that subagents exist in the same literal sense as tables and chairs. All they have to concede to is that some external agents behave *as if* their behavior is governed by separate subagents. The key premise of the view defended here is thus less controversial than one might think. Moreover, even if we insist on reading Plato's Theory of Subagents literally, there would still be cases in which the theory is relatively uncontroversial. Consider, for instance, games played by collective agents, such as the board of a large organization. The board is an external agent playing games against other collective, external agents, and each board member is an internal subagent of this collective agent. It is not unreasonable to think that decisions taken by a collective agent can sometimes be determined by internal games played by the subagents of the collective agent.

11.2 Indefinitely repeated internal Prisoner's Dilemmas

Before proceeding, it is helpful to recapitulate some basic facts about the Prisoner's Dilemma. In the generic (external) single-shot, two-person Prisoner's Dilemma illustrated in Figure 11.1 on the next page each player's outcomes rank as follows: $A > B > C > D$.[3]

External single-shot two-person Prisoner's Dilemmas sometimes occur in situations where we might not expect to find them. Consider the following example:

The Headlamps Game

You are driving along an empty highway late at night. Your headlights are in the high beam mode. You suddenly see another car coming towards you. Its headlamps are also in the high beam mode. You and the other driver know that you will never meet again. (Even if you did, you would not recognize the headlights you see as a car you have met before.) Both drivers

[3] We also assume that $B > (A + D) / 2$. This ensures that players cannot benefit more from alternating between cooperative and non-cooperative moves in iterated games, compared to mutual cooperation.

		Col	
		Cooperate	Do not
Row	Cooperate	B, B	D, A
	Do not	A, D	C, C

Figure 11.1 The Prisoner's Dilemma

have to decide simultaneously whether to dip the lights. The four possible outcomes are as follows.

(A) You keep high beam and the other driver dips the lights.
(B) Both drivers dip the lights.
(C) Both drivers keep high beam.
(D) You dip the lights and the other driver keeps high beam.

The two drivers face an external single-shot Prisoner's Dilemma. Each driver will be better off by *not* dipping his lights, no matter what the other driver decides to do, so $A > B > C > D$. Despite this, many drivers around the world do actually dip their lights, even when they meet drivers they know they will never meet again (and even when they know that there is no risk of being caught by the police or of being punished in some other way).

Empirical studies indicate that about 50 percent of laypeople cooperate when confronted with a single-shot, two-player Prisoner's Dilemma.[4] A possible explanation is that various social norms influence our behavior.[5] Social norms summarize an expectation about how other people will behave, which a rational player will take into account when playing the game. In many cases, the rational response to such an expectation is to reconstitute the norm by acting in accordance with it, which in turn makes it rational for others to also take this expectation into account and act in accordance with the norm.

However, even if we can explain why certain norms have emerged, and sometimes face situations in which it is rational to take them into account, a fundamental normative challenge still remains to be addressed: Why should the individual who faces a single-shot Prisoner's Dilemma care about these

[4] See Camerer (2003).

[5] The literature on this topic is very extensive. For an influential and stringent account, see Bicchieri (2006).

norms when the mechanisms that support the norm are not in place? In the Headlamps Game, the non-cooperative act dominates the cooperative one, so no matter what you expect the opponent to do, you will be better off by not cooperating. Why not just ignore whatever norms you normally care about and perform the dominant non-cooperative alternative?

A possible response could be that the norms in question *alter* the agent's preference ordering. But this would entail that the game is no longer a Prisoner's Dilemma, which makes the notion of norms somewhat uninteresting in the present context. If a game thought to be a Prisoner's Dilemma turns out not to have this structure, because the agents' all-things-considered preferences are different from what they appeared to be, this tells us nothing about whether it is rational to cooperate in the single-shot Prisoner's Dilemma.[6] The normative question that is the focus of this chapter remains unanswered: Is it ever rational to cooperate in a single-shot Prisoner's Dilemma?

Another and more sophisticated response is to introduce the concept of *indirect reciprocity*.[7] Even if A and B play the Prisoner's Dilemma only once, B has to take into account that B will play against other players in the future. Therefore, if B does not cooperate with A, and A informs C and D about this, B will quickly gain a bad reputation in the community. Whoever B meets in the future will know that B is not likely to cooperate and will therefore adjust her strategy to this. However, if B instead cooperates with A, and A passes on this information to others, B will gain a good reputation in the community and future opponents will know that B is likely to cooperate. The same applies to all other players: By not cooperating in a one-shot game they are likely to ruin their reputation, which will harm them in the long run. The upshot is that selfish players can benefit more in the long run by cooperating in one-shot games than by not cooperating, because this often makes it more likely that other players (who also seek to protect their reputation) will cooperate with them. So it may very well be rational to cooperate in the one-shot game if one takes into account how one's reputation in the community will be affected.[8]

[6] Cf. Binmore's contribution to this volume.

[7] See Nowak and Sigmund (2005) for an overview of the literature on indirect reciprocity and community reinforcement.

[8] A substantial amount of research into theories of indirect reciprocity have been based on computer simulations. Researchers have tried to determine under exactly what conditions indirect reciprocity is the successful strategy for members of a community who play a series of one-shot games against new opponents in the community. The review article by Nowak and Sigmund (2005) gives a good overview.

Although the literature on indirect reciprocity is interesting and helps us to shed light on many social and biological processes, it is worth keeping in mind that indirect reciprocity is not a good reason for cooperating in the Headlamps Game. When you meet other cars in the dark you know that most of them will play the cooperative strategy, but because it is dark they will not be able to identify you and gossip about your non-cooperative behavior. All they will learn when you refuse to cooperate is that there is exactly one non-cooperative player on the roads. This will not affect their behavior. Moreover, if considerations of indirect reciprocity were to actually change the drivers' behavior they would no longer be playing a one-shot Prisoner's Dilemma. As explained above, a game in which other considerations and values are introduced might be interesting for many reasons, but it is no longer a one-shot Prisoner's Dilemma.

Compare the external single-shot version of the Headlamps Game with the following repeated version of the same game.

The Headlamps Game on a small island

You live on a small island, on which there is only one other car (which is always driven by the same person). Every time you meet a car in the dark, you know that you are playing against the same opponent. Although the actual number of rounds may well be finite, there is no pre-determined last round of the game, meaning that we cannot solve the game by reasoning backwards.[9] Both drivers must base their decisions about what to do on their expectations about the future. Under reasonable empirical assumptions, each player's expected future gain of cooperating will outweigh the short-term benefit of the non-cooperative strategy. The upshot is that the two drivers will quickly learn to cooperate, because it is in their own best interest to do so.

The repeated Headlamps Game illustrates a general and important insight about the Prisoner's Dilemma: In *indefinitely* repeated versions of the game, in which the probability is sufficiently high in each new round that that round is not the last, each player's expected future gain of cooperating will outweigh the short-term benefits of the non-cooperative move. According to the Folk

[9] If the number of rounds to be played is known at the outset, the backwards induction argument tells us that if the players know that they will play the game n times, then they will have no reason to cooperate in the n:th round. Therefore, since the players know that they will not cooperate in the n:th round, it is irrational to cooperate in round $n - 1$, and so on and so forth, for all rounds up to and including the first round. Some of the technical assumptions of the backwards induction argument are controversial.

Theorems for indefinitely repeated games (this is a set of theorems that cover slightly different types of games that were known and discussed by game theorists long before they were formally proven), stable cooperation can arise if players play strategies that are sufficiently reciprocal, i.e. if each player cooperates with a high probability given that the other player did so in the previous round, but otherwise not.[10] Therefore, "in the shadow of the future," we can expect rational players to cooperate in indefinitely repeated Prisoner's Dilemmas.

Let us now put a few things together. First, we know that in an ordinary external single-shot Prisoner's Dilemma the only rationally permissible alternative is to play the non-cooperative strategy, because it dominates the cooperative strategy. Second, we know that in indefinitely repeated versions of the game it will, under some reasonable empirical assumptions, be rational to cooperate. This naturally leads to the question whether a set of alternative strategies can *simultaneously* figure as alternative strategies in an external single-shot Prisoner's Dilemma *and* in another indefinitely repeated internal version of the game.

Enter the Platonic Theory of Subagents. When Anne and Bob play an external single-shot Prisoner's Dilemma, Anne's subagents may at the same time be playing another internal game that determines her decision in the external game. The internal game played by Anne's subagents can be described as an indefinitely repeated game (at least if Anne believes that she is not about to die in the near future). The same is true of Bob and his subagents.

This very rough sketch indicates that there may exist a class of external single-shot Prisoner's Dilemmas in which agents cooperate for what appears to be the same reason as rational players cooperate in some indefinitely repeated versions of the game. The next section examines the details of this argument.

11.3 My argument

Given that we accept the Platonic Theory of Subagents, we could expect some internal games played by our subagents to be similar in structure to external Prisoner's Dilemmas played by ordinary agents. The Prisoner's Dilemma can

[10] For a useful introduction and overview of the Folk Theorems, see chapter 6 in Hargreaves-Heap and Varoufakis (2004). Two of the most well-known examples of reciprocal strategies are grim trigger and Tit-for-Tat.

thus arise in a place that has often been overlooked by game theorists: inside ourselves. For instance, if you are about to buy a new car, several subagents, each with a separate preference ordering, may influence your choice. Each subagent may rank attributes such as safety, fuel economy, and elegance differently.[11] Which car you buy depends on the equilibrium reached in the game played by your subagents, and some of these internal games will be similar in structure to the indefinitely repeated Headlamps Game discussed above.

Although it is interesting in its own right to point out that a Prisoner's Dilemma can arise "within" the agent, this insight is in itself no reason for revising the orthodox analysis of single-shot external Prisoner's Dilemmas. However, as explained above, I propose that if we combine this idea with the thought that some internal games are repeated indefinitely, we obtain a powerful argument for thinking that it is sometimes rational to cooperate in an external single-shot Prisoner's Dilemma. (Note that if a game is repeated indefinitely many times it does not follow that it is repeated infinitely many times. This just means that there is no pre-determined last round; in each round there is a non-zero probability that that round is not the last.[12])

The key premise of my argument is that there is a sense in which single-shot external Prisoner's Dilemmas can be thought of as being parts of other, indefinitely repeated internal Prisoner's Dilemmas. When such single-shot external Prisoner's Dilemmas figure in other indefinitely repeated internal Prisoner's Dilemmas, it is often rational to cooperate. In the shadow of the future, each subagent benefits more from cooperating than from defecting, because the subagents will meet each other again in future rounds of the internal game. In order to explain in greater detail how this argument works, it is helpful to consider an example.

Selfish vs. altruistic donations

You are about to decide whether to make a large donation to your Alma Mater. You also have to decide whether to brag to others about the donation. What you eventually decide to do will be determined by the preference orderings of your subagents. To keep the example simple, we assume that only two subagents influence your decision: an egocentric and an altruistic one. Depending on the strategies played by the subagents

[11] This example is discussed in detail in Kavka (1991).

[12] This assumption blocks the backwards-induction argument, which presupposes that the player knows before the last round actually occurs that that round will be the last.

Table 11.1 Altruistic and egocentric subagents

	Altruistic subagent	Egocentric subagent
(A)	$d \wedge \neg b$	$\neg d \wedge b$
(B)	$d \wedge b$	$d \wedge b$
(C)	$\neg d \wedge \neg b$	$\neg d \wedge \neg b$
(D)	$\neg d \wedge b$	$d \wedge \neg b$

(who control different parts of the decision-making process) you will end up performing exactly one of the following four acts:

$d \wedge b$ Donate and brag about the donation.
$d \wedge \neg b$ Donate and do not brag about the donation.
$\neg d \wedge \neg b$ Do not donate and do not brag about the donation.
$\neg d \wedge b$ Do not donate and brag about the donation.

Your two subagents rank the four acts differently. In order to facilitate comparisons with the generic Prisoner's Dilemma in Figure 11.1 it is helpful to label the four alternatives with capital letters. In Table 11.1, the four alternatives are listed from the best to the worst, i.e. $A > B > C > D$.

Although it does not matter *why* the subagents have the preferences they have, it is not difficult to construct a story that could explain this. Imagine, for instance, that the reason why the altruistic subagent has the preference ordering listed in Table 11.1 is that she assigns a high weight to the fact that a donation is made but cares less about bragging. The egocentric subagent assigns a lot of weight to the fact that the agent brags, but cares less about the donation. The two subagents thus agree on the ranking of $d \wedge b$ and $\neg d \wedge \neg b$, but they do so for different reasons.

Note that the preference orderings of the two subagents fulfill the conditions of a Prisoner's Dilemma. We can, if we so wish, describe the altruistic subagent as a moral decision-maker concerned with the wellbeing of others and the egocentric subagent as a decision-maker concerned primarily with her own prestige and reputation.

Let us imagine that your egocentric and altruistic subagents face the internal game described above many times and that they both believe in every round that there is some non-zero probability that that round is not the last. Both subagents then have to think ahead and take the expected future gain of cooperation into account, and then compare this to the short-term benefit of the non-cooperative strategy. Given some reasonable assumptions about each opponent's strategy in this indefinitely repeated game, it is rational for the two subagents to cooperate, meaning that they will reach a stable Nash equilibrium in which the external agent (you) make a donation and brag about it.

If we were to state the argument proposed here in its most general form, the technical details would inevitably become very complex. There is a risk that we would fail to see the wood for the trees. So in this chapter, the best way forward is to consider the simple case in which both subagents play the very simple *grim trigger strategy*. By definition, a player playing the grim trigger strategy cooperates in the initial round of the game, and in every future round, as long as the other player also cooperates. However, as soon as the other subagent defects, players playing the grim trigger strategy will defect in *all* future rounds of the game, no matter what the other player does. A non-cooperative move will never be forgiven.

Suppose that each subagent knows that the opponent plays the grim trigger strategy. Under what conditions will it then be rational for the subagents to cooperate? Without loss of generality, we can set the utility of the outcome in which both players do not cooperate to zero (because utility is measured on an interval scale). So in Figure 11.2 it holds that C=0 and A> B> 0 > D.

We know that if both subagents cooperate in every round, and the probability is p in each round that that round is not the last, the expected utility of cooperating is $B + pB + p^2B + p^3B\ldots = B/(1 - p)$ for each subagent. Moreover, a subagent who does not cooperate in the current round will get A units of utility in this round and 0 in all future rounds (because the opponent plays the grim trigger strategy). By putting these two facts together, it can be easily verified that it is rational for each subagent to cooperate in the current round if and only if $B/(1 - p) \geq A$, which entails that each subagent will cooperate as along as $p \geq 1 - (B/A)$. So, for example, if the utility of $d \wedge \neg b$ is 2 units and that of $d \wedge b$ is 1 unit, it is rational for each subagent to cooperate as long as the probability is at least ½ that the current round is not the last.

		Col	
		Cooperate	Do not
Row	Cooperate	B, B	D, A
	Do not	A, D	0, 0

Figure 11.2 The grim trigger strategy

If the subagents do not play the grim trigger strategy but play some other reciprocal strategy (such as Tit-for-Tat, with some probability of forgiving non-cooperative moves) then the calculations will be somewhat different. But the general message still holds: Rational subagents should cooperate because they run a risk of being punished in the future if they don't. The argument spelled out here is in fact rather insensitive to which particular strategy one thinks a rational agent would play in the indefinitely repeated Prisoner's Dilemma. As explained in Section 11.2, the Folk Theorems teach us that cooperation will always be rational, given that the probability is sufficiently high that the current round is not the last and given that all subagents play sufficiently reciprocal strategies.

The example sketched above is just one of many illustrations of single-shot Prisoner's Dilemmas in which it is rational to cooperate. Consider, for instance, the single-shot Headlamps Game discussed earlier. Imagine that each external agent has two internal subagents and that the first subagent is an altruistic subagent whose primary desire is to dip the lights whenever she meets another car. Her second, less weighty desire is to not exceed the speed limit. The second subagent is more selfish. Her primary desire is to drive as fast as possible (and exceed the speed limit). Her second, less weighty desire is to not get distracted by dipping the lights. By combing the two choices, dipping the lights (d) or not (¬d), and exceeding the speed limit (e) or not (¬e), we see that the preferences of the two subagents determine which of the four possible alternatives the driver will eventually perform. Consider Table 11.2, which is analogous to Table 11.1.

If the game described in Table 11.2 is repeated indefinitely many times, with sufficiently high probability, then the subagents of the first driver will cooperate, i.e. play $d \wedge e$. If we believe that the behavior of the second driver is also determined by the preferences of two (or more) subagents, then those subagents will also cooperate. Hence, both drivers will dip their lights and exceed the speed limit, despite the fact that they play a single-shot Prisoner's Dilemma.

Table 11.2 The indefinitely repeated Headlamps Game

	Subagent 1	Subagent 2
(A)	$d \wedge \neg e$	$\neg d \wedge e$
(B)	$d \wedge e$	$d \wedge e$
(C)	$\neg d \wedge \neg e$	$\neg d \wedge \neg e$
(D)	$\neg d \wedge e$	$d \wedge \neg e$

Clearly, the two examples outlined here have the same structure. The general recipe for generating further examples is simple: (1) Take a game in which the choice of each external agent is determined by a game played by two internal subagents. (2) Ensure that for each external player of the single-shot game, there is one altruistic and one selfish subagent. (3) Then, given reasonable assumptions about the preferences and beliefs of these subagents, the subagents face an indefinitely repeated Prisoner's Dilemma. (4) Because the selfish and altruistic subagents have reason to believe that they will meet each other in future rounds of the game, the two subagents both benefit from cooperating.

11.4 Not a Prisoner's Dilemma?

At this point it could be objected that the external game played by the drivers in the Headlamps Game is not a genuine Prisoner's Dilemma, because the internal games played by the subagents *change* the preferences of the external agents.[13] As Binmore puts it in his contribution to this volume:

> Critics of the orthodox analysis focus not on the game itself, but on the various stories used to introduce the game. They then look for a way to retell the story that makes it rational to cooperate... If successful, the new story necessarily leads to a new game in which it is indeed a Nash equilibrium for both players to [cooperate]. (Binmore, this volume, p. 23)

The best response to Binmore's objection is to stress that the structure of the original single-shot game has not changed. The external agents still have the preferences they have, and those preferences constitute a single-shot Prisoner's Dilemma. However, as Gauthier argues in his contribution to this volume, the preference of what I call the external agent is not always revealed in her choice behavior. The external agent *prefers* not to cooperate in the single-shot Prisoner's Dilemma, but the *rational choice* is nevertheless to cooperate, because of the structure of the internal game played by the subagents.

In the revised version of the single-shot Headlamps Game discussed above, the two subagents dip their lights and exceed the speed limit in the internal game. But this does not entail that the agents' preferences in the external game have changed. The external agents still prefer to not dip their lights in

[13] I would like to thank Paul Weirich for raising this objection with me.

the single-shot Headlamps Game. Under the assumption that our choices are outcomes of games played by internal subagents, preferences and choices sometimes come apart. Consider Plato's thirsty man mentioned in the introduction. The thirsty man prefers to drink. But it does not follow that the rational choice for the external agent is to drink, because the rational choice is an outcome of a game played by several internal subagents. What makes it rational for the external agent to refrain from *choosing* according to his preference is the structure of the game played by these internal subagents.

It is worth pointing out that the Platonic account of choice and preference is incompatible with revealed preference theory. According to the Platonic account, the fact that an option is chosen does not entail that it is preferred. There is an extensive and growing literature that questions the truth of revealed preference theory.[14] Revealed preference theory is no longer something that is universally accepted by all game theorists.

11.5 The Strong and the Weak Interpretations

Let us consider two further objections. First, critics could ask whether it really makes sense to claim that people are governed by altruistic and selfish subagents. What arguments, if any, can be offered for this psychological claim? The second objection has to do with the generalizability of the argument. Even if it is *possible* that situations of the type described here can arise, it remains to determine how common they are. If it only happens under exceptional circumstances that rational agents governed by two or more subagents have reason to cooperate, it seems that my point is of little general interest.

Let us begin with the first objection. How plausible is it to claim that an external agent's decision to cooperate in the Prisoner's Dilemma is an outcome of a game played by two or more internal subagents? The answer of course depends on how we interpret the Platonic Theory of Subagents. At least two fundamentally different interpretations are possible. I shall refer to these as the *Strong* and the *Weak Interpretations.*

According to the Strong Interpretation, subagents exist in more or less the same sense as tables and chairs. Ordinary people literally have subagents in their heads. I am an agent, and somewhere within me there exist two or more subagents whose decisions jointly determine what I do. There is no fundamental difference between the internal game played by my subagents and, say, a game of chess played by two Grand Masters.

[14] See, for instance, Gauthier's and Hausman's contributions to this volume.

The obvious weakness of the Strong Interpretation is that there seems to be very few controlled scientific studies that support it.[15] The consensus view among contemporary psychologists is that Plato was wrong, at least if we interpret him along the lines suggested here. The Platonic Theory of Subagents is perhaps a useful metaphor that captures something we all tend to experience from time to time. But it is not a claim that can be interpreted literally and supported by scientific evidence.

Let us now consider the Weak Interpretation. According to this interpretation, the Platonic Theory of Subagents is just an analytic tool. We should not believe that people are *actually* made up of separate subagents. The point is rather that people behave *as if* their behavior is guided by internal subagents. Kavka explicitly defends the Weak Interpretation:

> I suggest we explore some of the implications of treating individual suborderings, desires, criteria, or dimensions of evaluation *as though* they were represented by distinct subagents (within the individual) who seek to achieve satisfaction of the desire (criterion, etc.). I do *not* claim that we are in fact composed of multiple distinct selves, each of which forms an integrated unit over time and has separate dispositions or values from the other selves of the same individual. (Kavka 1991: 148)

The main attraction of the Weak Interpretation is that it is less demanding than the strong one from a metaphysical point of view. No matter whether we have reason to think that subagents really exist, we could nevertheless reason *as if* they exist. Anyone who is familiar with Bayesian decision theory will recognize this argumentative strategy:[16] Bayesian decision theory is a theory of rational choice for individuals confronted with a set of uncertain prospects. According to this theory, preferences over uncertain prospects must fulfill certain structural conditions in order to be rational. Given that these structural constrains are met, it can be shown that a rational agent behaves *as if* his or her choices were governed by the principle of maximizing subjective expected utility.

In order to assess the plausibility of the Weak Interpretation it is helpful to discuss the analogy with Bayesian decision theory a bit further. I believe there are important similarities, but also a crucial dissimilarity.

[15] It is worth pointing out that Kavka (1991) claims that there is at least one scientific study that supports what I call the Platonic Theory of Subagents.

[16] See, e.g., Ramsey (1926), Savage (1954), and Jeffrey (1983).

Note that Bayesian decision theorists do not claim that people actually have utilities and subjective probabilities in their heads. Utilities and subjective probabilities are abstract entities constructed for predicting and explaining people's behavior when confronted with a certain type of decisions. The as-if clause in the Weak Interpretation of the Platonic theory is meant to function in the same way as the corresponding clause in Bayesian decision theory. Advocates of the Weak Interpretation do not ask us to believe that people have subagents in their heads. On the contrary, subagents are fictional entities we use for rationalizing people's behavior.

The main advantage of the Weak Interpretation is that it is less demanding from a metaphysical point of view than the Strong Interpretation. However, it is not clear, at least not at this stage, that the Weak Interpretation can really support the claim that it is rational to cooperate in single-shot Prisoner's Dilemmas. In order to see this, it is important to distinguish between descriptive and normative applications of game and decision theory.

First, consider descriptive interpretations of Bayesian decision theory. Although one can of course question whether people actually behave in accordance with the axioms of the theory (there is plenty of empirical evidence to the contrary), the as-if clause is rather unproblematic in descriptive interpretations.[17] No matter whether people actually have subjective probabilities and utilities in their heads, it can be helpful from a predictive as well as from an explanatory point of view to ascribe such entities to agents. As long as the decision theorist is able to elicit accurate predictions and fruitful explanations there is little to worry about.

However, when we turn to normative interpretations of the Bayesian theory it is far from evident that the as-if clause gives the normative decision theorist what she wants. The problem is that normative versions of Bayesian decision theory put the cart before the horse, meaning that the theory is not able to offer the agent sufficient action guidance. F. P. Ramsey, the inventor of Bayesian decision theory, was aware of this problem. In a brief note written two years after "Truth and Probability," Ramsey remarks that

> sometimes the [probability] number is used itself in making a practical decision. How? I want to say in accordance with the law of mathematical expectation; but I cannot do this, for we could only use that rule if we had measured goods and bads. (Ramsey 1928/1931: 256)

[17] For a brief summary of the descriptive literature, see Peterson (2009: chapter 14).

Let me try to unpack Ramsey's point. The reason why he cannot advise a rational agent to apply the principle of maximizing expected utility for making decisions is that the agent's preferences over the set of available alternatives is used for *defining* the notion of utility (and subjective probability). The "goods" and "bads" are derived from the agent's preferences over all available prospects and this preference order is supposed to be complete. Hence, Ramsey's rational agent is supposed to know already from the beginning whether he prefers one alternative prospect (that is, one good or bad) to another, meaning that for the ideal agent no action guiding information can be elicited from the theory. The completeness axiom entails that the rational agent already knows what to do, since the set of entities the agent is supposed to have complete preferences over includes all the acts she can choose to perform. This is why Ramsey writes that he cannot say what he wants to say.

For non-ideal agents, whose preferences are either incomplete or violate at least one of the other preference axioms, it is of course true that Ramsey's theory (as well as the theories proposed by other Bayesians) can be action guiding in an indirect sense. If a non-ideal agent discovers that her current preferences are incomplete, or violate some of the other axioms, then the agent should revise her preferences such that all the axioms are fulfilled. However, the theory tells us nothing about *how* the preferences should be revised. Any revision that yields a set of preferences that obeys the axioms will do.

Let me try to relate these insights about as-if reasoning in Bayesian decision theory to the Prisoner's Dilemma. First, note that as long as we use game theory for descriptive purposes, such as for explaining and predicting how people react when confronted with the Prisoner's Dilemma, it seems to be no more problematic to refer to hypothetical subagents than to hypothetical utility and probability functions. Given that the predictions and explanations are accurate, it is legitimate to claim that people behave *as if* their decisions are governed by something that might in the end not exist.

However, when we turn to normative issues in game theory, it is yet unclear whether it is sufficient to rely on as-if reasoning. What is the normative relevance of the fact that we can reason as if our decisions were governed by subagents if we suspect that no such subagents exist? To put it briefly, the worry is that the normative relevance of as-if reasoning in game theory is no greater than the normative relevance of as-if reasoning in Bayesian decision theory.

In order to steer clear of this objection, it is important to observe that there is in fact a crucial difference between as-if reasoning in game theory and

Bayesian decision theory. In the analysis of the single-shot Prisoner's Dilemma proposed here, the key idea is that ordinary agents like you and me should reason as if our decisions were governed by a set of subagents playing an indefinitely repeated Prisoner's Dilemma. The reason why we should reason as if such subagents determine our choices is that this is a good *descriptive* account of how humans function. It is not a normative claim about what makes a set of preferences rational, or about how non-ideal agents should revise their preferences. The point is merely that we should take into account how agents function when we formulate normative principles for how a rational agent should reason when confronted with the Prisoner's Dilemma.

As noted earlier, about 50 percent of ordinary people without training in game theory cooperate with their opponent when confronted with the single-shot Prisoner's Dilemma.[18] By reasoning *as if* our decisions were determined by games played by subagents, we can explain why this is so. Hence, the theory of subagents is not unreasonable if interpreted as a descriptive as-if claim.

Once we believe that the as-if account of ourselves is descriptively accurate, this makes it plausible to defend the following normative claim: If an agent who plays the Prisoner's Dilemma can be accurately described as someone who believes that her subagents will start to punish each other if she does not cooperate now (that is, if the agent can be described *as if* this was the case), then the agent should take this insight into account when deciding what to do in a single-shot Prisoner's Dilemma. Therefore, the Weak Interpretation is sufficiently strong for warranting the claim that it is rational to cooperate in single-shot Prisoner's Dilemmas.

The upshot of all this is that the Weak Interpretation is more attractive than the strong one. We should reason *as if* the Platonic Theory of Subagents is correct, without claiming that such subagents actually exist. A similar argumentative strategy is widely accepted in Bayesian decision theory, where it works well for descriptive purposes. (All I claim is that this argumentative strategy is coherent from a conceptual and philosophical point of view. The fact that it may very well be an inaccurate descriptive account is irrelevant in the present context.) The as-if approach to normative decision theory is, however, of limited value since it offers us no or little action guidance. The fully rational Bayesian agent already knows what to do. And non-ideal agents

[18] See Section 11.2 and Camerer (2003).

can merely use the theory for revising their preferences such that they become compatible with the structural axioms of rational preferences proposed by Bayesians.

11.6 Possible generalizations

As noted in the introduction, there is at least one type of game in which the claim that external agents consists of internal subagents is uncontroversial irrespective of how it is interpreted. The games I have in mind are games played by collective agents, such as firms and other organizations. For such games, even the Strong Interpretation seems applicable. Imagine, for instance, that the board of a large company is thinking of doing business with a foreign company they have never worked with in the past. If the two companies manage to negotiate a deal, they will only do business for a short period of time and they will never interact again in the future. Suppose that the game played by the two companies is (for reasons we need not worry about here) a single-shot Prisoner's Dilemma. Would it be rational for the two boards to cooperate?

If we generalize the Platonic Theory of Subagents from ordinary humans to collective agents (as Plato famously did himself in the *Republic*) we see that it may be rational for the external players to cooperate in this single-shot Prisoner's Dilemma. The board of each company can be viewed as a collective external agent comprising several internal subagents. In each board meeting the members of the board know that there is a sufficiently high probability that that board meeting is not the last. Therefore, it is rational for them to take into account how other board members will behave in the future. The upshot is that each board member has to consider how other board members would react if he or she were to play a non-cooperative strategy.

It is not crucial to the argument to determine precisely what range of real-life cases can be covered by the Strong Interpretation. As emphasized in Section 11.5, the Weak Interpretation is sufficiently strong for justifying the normative claim about single-shot Prisoner's Dilemmas that is central to this chapter. So the strength of the argument defended here primarily depends on how large the range of cases is that is covered by the Weak Interpretation.

Unsurprisingly, it is not difficult to identify at least *some* single-shot Prisoner's Dilemmas to which the argument for cooperation does not apply. Imagine, for instance, that Alice and Bob play a single-shot Prisoner's Dilemma and that they know that this is the last decision they will make before they die. Once they have decided what to do in the current round, a pandemic disease will kill them, which they know. In this situation it would clearly not be rational for Alice and Bob to cooperate, simply because their subagents know that they will never get an opportunity to play against their opponents again in the future.

The insight that the argument for playing cooperative strategies merely applies to *some* single-shot Prisoner's Dilemmas might make us doubt the scope of the argument. Perhaps the fraction of single-shot Prisoner's Dilemmas to which the argument applies is very small or even minuscule? My response to this worry consists of two points. First, it is arguably interesting in its own right to point out that there is a least *some* single-shot Prisoner's Dilemmas in which it is rational to cooperate. The consensus view in the literature is that there is *no* such game in which it is rational to cooperate.

My second point is less defensive. Here is what I have to say: Although it is true that not every external single-shot Prisoner's Dilemma can be reconstructed as some game in which the agents play another set of internal Prisoner's Dilemmas guided by the preference orderings of their subagents, this does not entail that the argument outlined here fails. On the contrary, the key idea of the argument can be applied to a much broader set of games. In order to see this, note that it can be rational for the external agents playing the single-shot Prisoner's Dilemma to cooperate even if the internal games played by the subagents are not Prisoner's Dilemmas. All that matters for the argument to go through is that the internal games are ones in which it is rational for the subagents to cooperate. Needless to say, there are numerous other games, repeated as well as non-repeated ones, in which it is rational for subagents to cooperate. The key questions that determine the scope of the argument are thus the following: (1) Can every external single-shot Prisoner's Dilemma be reconstructed as a game played by two or more internal subagents? (2) If so, is it rational for the subagents to cooperate in those internal games?

As we have seen above, the answer to the second question varies from case to case. There is a wide range of cases in which it is rational for two or more subagents to cooperate, but this is not always the case. There are exceptions.

But what about the first question: Can every single-shot external Prisoner's Dilemma really be reconstructed as a game played by two or more subagents? The answer is yes. From a technical point of view, it is trivial that *every* game can be reconstructed along the lines proposed here. A game is essentially a list of preference orderings expressed by agents over a set of possible outcomes. It takes little reflection to see that it is always possible to construct *some* game, which we may refer to as an internal game, that has the alternative acts corresponding to those preference orderings as outcomes.

A final worry about the argument arises even if we think it is indeed true that some single-shot external Prisoner's Dilemmas can be represented as a game in which the external agent's behavior is governed by a set of internal games played by subagents. The worry is that nothing seems to guarantee that this representation is unique. There might very well be other ways of representing the game, according to which it would not be rational to cooperate. What should a rational agent do if it is rational to cooperate according to some representation of the game, but not according to others? The best response is to point out that although this is an issue that is interesting and worth discussing, this is also a very general objection that takes us far beyond the scope of the present chapter. Very broadly speaking, nothing seems to exclude that in some situations one and the same set of facts about a situation can be described equally well by different games. This insight is not unique for the current discussion. In fact, game theorists have had surprisingly little to say about how we should determine which game someone is actually playing. It is one thing to view a game as a formal object and study its technical properties, but a completely different task to determine how well such a formalization represents the real-world phenomena we are ultimately interested in.

If the formalization I propose is plausible, this may have important implications for how we should address single-shot Prisoner's Dilemmas in society. The good news is that we would then actually benefit more from cooperating with each other than has been previously recognized.

12 Prisoner's Dilemmas, intergenerational asymmetry, and climate change ethics

Douglas MacLean

12.1 Introduction

Climate change is happening, and human activity is the cause. What are we going to do about it? Bill McKibben calls this "the most important question that there ever was."[1] It is a practical question, but it also raises some philosophical issues. My goal in this chapter is to examine these philosophical issues. They include matters of self-interest and rational choice, and they include moral issues about the value of nature, the nature of human values, and difficult issues about international justice and intergenerational morality.

A number of writers have pointed out that climate change has remarkable spatial and temporal characteristics. It is a global problem and a paradigm of what game theorists call the Prisoner's Dilemma. The fundamental characteristic of a Prisoner's Dilemma is that if each agent successfully pursues its rational best interests, the result is collectively worse for each of them than some other possible result. I will discuss this issue in Section 12.3 below.

The temporal problem is an illustration of intergenerational asymmetry, or what some philosophers call the "tyranny of the present."[2] In each generation, people must decide whether they will collectively accept some costs in order to reduce greater harms and costs in the future or continue to pursue short-term gains and pass the problem on to the next generation. Because future people do not exist, they cannot bargain, reciprocate, compensate, reward, or punish us for what we do. I will discuss the implications of this asymmetry in intergenerational morality in Sections 12.4 and 12.5.

For helpful comments and suggestions on an earlier draft of this paper, I am indebted to Martin Peterson and Susan Wolf.

[1] The quote comes from a tribute McKibben gave at a celebration of the retirement of James Hansen, one of the pioneers of climate modeling and an early prophet of climate change. See Gillis (2013).

[2] See Gardiner (2011).

Both of these philosophical problems are familiar, and I will have little to add to the technical analysis of them. Once one understands and accepts the central facts about climate change, moreover, it is easy enough to formulate principles that tell us what we must do to avoid causing the worst harms and to mitigate the future harms we have already caused. But, to expand the normal meaning of the moral dictum that "'Ought' implies 'can,'" I will insist that plausible moral principles must be connected to reasons for action that express our values in a way that can plausibly motivate us to act. This makes the problem of finding adequate moral principles more complicated. Thus, I will explain in Section 12.7 how the spatial and temporal dimensions of climate change interact to make solving this problem more difficult. This will involve some further comments about the relation between moral values and political institutions.

I will, in effect, be explaining and defending a remark about climate change that Vice President Al Gore made nearly two decades ago: "The minimum that is necessary far exceeds the maximum that is politically feasible."[3] This is a pessimistic conclusion, but not a hopeless one. In the end, I believe the main reason it is so difficult to respond constructively to climate change, and what explains our failure to act, has to do with the nature of our democratic politics that so favor near-term and local goals over more distant ones. This is a bias that not only makes it difficult to support policies that reflect our deeper values in responding to climate change but also distorts the meaning of democracy.

12.2 Knowledge, predictions, and "Natural Hazards"

A consensus exists among scientists about the causes of climate change and about some of its effects. Climate science predicts some consequences under different scenarios, although uncertainty inevitably surrounds the timing and magnitude of these consequences. The authoritative source of information about climate science is the Intergovernmental Panel on Climate Change (IPCC), a group of hundreds of scientists and government officials representing 195 countries. The IPCC was organized by the United Nations Environmental Program and the World Meteorological Organization in 1988 to issue periodical reports on the state of scientific knowledge about climate change and its impacts. Its Fifth Assessment Report, issued in stages in 2013–14,

[3] Vice President Gore made this remark in an interview in 1992 with Bill McKibben. It is quoted in McKibben (2001).

leaves no doubt that anthropogenic climate change is happening, and the impacts are already being felt.[4] These conclusions are echoed by other reports on the state of climate science. The US government's third National Climate Assessment (NCA), issued in May 2014, explains in its Overview that

> Over recent decades, climate science has advanced significantly. Increased scrutiny has led to increased certainty that we are now seeing impacts associated with human-induced climate change. With each passing year, the accumulating evidence further expands our understanding and extends the record of observed trends in temperature, precipitation, sea level, ice mass, and many other variables recorded by a variety of measuring systems and analyzed by independent research groups from around the world. It is notable that as these data records have grown longer and climate models have become more comprehensive, earlier predictions have largely been confirmed. The only real surprises have been that some changes, such as sea level rise and Arctic sea ice decline, have outpaced earlier predictions.[5]

Of course, some people, including some powerful politicians in the US, refuse to acknowledge these facts, but their denials can no longer be taken seriously. To refuse to believe today that future climate disasters will be more likely if we fail to act soon to reduce greenhouse gas emissions is no more credible than to deny that life evolves. The mere fact that some people claim that the earth is flat or deny that smoking harms human health is not a morally excusable reason to believe them or to remain agnostic. One needs to examine the evidence to determine what is credible, and we should hold each other morally responsible for the consequences of denying uncomfortable facts and remaining culpably ignorant in the face of overwhelming evidence. Even the arch-pragmatist William James insisted that one should no longer continue to hold a belief in the face of convincing evidence to the contrary, and the evidence of climate change is now clear and unambiguous.[6]

The basic facts are well known. Humans emit greenhouse gases, primarily CO_2, as a result of burning fossil fuels. CO_2 remains in the atmosphere for decades and centuries. Other natural processes also release CO_2 and other greenhouse gases but at a rate that keeps the overall concentration of gases in the atmosphere more or less constant, at least over the past several millennia.

[4] Intergovernmental Panel on Climate Change, Fifth Assessment Report, www.ipcc.ch/report/ar5/

[5] US Global Change Research Program (2014). [6] See James (1896).

Since the industrialized world came to rely on fossil fuels as the engine of progress beginning in the nineteenth century, however, human activity has increased the concentration of greenhouse gases in the atmosphere. Climate models that simulate how natural factors alone influence the climate system "yield little warming, or even a slight cooling, over the 20th century. Only when models include human influences on the composition of the atmosphere are the resulting temperature changes consistent with observed changes."[7]

Greenhouse gases absorb some of the sun's rays as they are reflected off the earth, which warms the atmosphere. This is the "greenhouse effect," first hypothesized by James Fourier in 1824 and confirmed toward the end of the nineteenth century by John Tyndall. In 1896, the Swedish chemist Svante Arrhenius suggested that burning fossil fuels could add to the greenhouse effect, but he thought this addition would be relatively benign compared to the warming caused by other natural forces. It wasn't until the mid-twentieth century that scientists began to notice that greenhouse gas emissions were increasing at an alarming rate, and atmospheric concentrations were reaching unprecedented levels, causing the average surface temperature of the earth to rise. No climate models existed then to predict the results of a rising temperature, and scientists were just beginning to drill and save ice core samples from Arctic ice sheets and mountain glaciers to get data on climate changes in the past. Although climate science was then in its early stages, the scientists' concerns were growing. In 1965, a report by President Johnson's Science Advisory Committee mentioned the greenhouse effect and warned that, "Man is unwittingly conducting a vast geophysical experiment."[8]

We now know that the concentration of greenhouse gases has increased 40 percent since 1800 and recently surpassed 400 parts per million, a level unprecedented in three million years. The average surface temperature on earth has increased by 1.4° F since 1900, and the thirty-year period between 1983–2013 is probably the warmest such period in over 800 years.

One certain consequence of global warming is a rise in sea level. Natural events, such as the processes that cause land masses to rise or sink, cause changes in sea level, but a warming climate also causes sea level rise because an increase in temperature causes water volume to expand and glaciers to melt. The overall rise in sea level since 1901 is about eight inches, and, according to the IPCC report, if greenhouse gases continue to increase at

[7] *Climate Change: Evidence & Causes* (2014), p. 5. [8] See Waert (2003).

their currently projected rate, the seas are predicted to rise by another 1.5–3 feet by 2100. Without major engineering projects to mitigate the effects of this increase, it will inundate the lowest lands and cause frequent severe flooding elsewhere, forcing many millions of people to abandon their homes and cities.

A rising average temperature also causes the earth's lower atmosphere to become warmer and moister, which creates more potential energy for storms and other severe weather events. The consequences, as one report explains, are that, "Consistent with theoretical expectations, heavy rainfall and snowfall events (which increase the risk of flooding) and heat waves are generally becoming more frequent… Basic physical understanding and model results suggest that the strongest hurricanes (when they occur) are likely to become more intense and possibly larger in a warmer, moister atmosphere over the oceans."[9] This prediction is now supported by observable evidence.

In light of these changes, we should revise our conception of natural disasters, which we tend to think of as "acts of God." Our responsibility for weather-related disasters now includes not only how we prepare for or respond to natural hazards but also how we have become partners with God in creating them.

The most recent IPCC report points out that some of the consequences of climate change are likely to be beneficial, especially in northern regions of the globe. Much depends on the rate of change and the ability of various plant and animal species to adapt. Most experts agree, however, that the negative effects will vastly outweigh the positive ones, and the worst of the negative effects will fall within tropical regions and affect millions of the poorest people on earth, who have both contributed least to the problem and are also the worst prepared societies to respond to the consequences.

Unless we change our behavior and vastly reduce our current rate of greenhouse gas emissions – that is, unless we move away from the "business-as-usual" scenario, as it is called – scientists worry that we will cause temperatures to rise more than 2° C (3.6° F) above pre-industrial levels before the end of this century. This amount of warming has been proposed as a threshold of our capacity effectively to adapt. The IPCC report, for example, expresses scientists' fears that "exceeding that level could produce drastic effects, such as the collapse of ice sheets, a rapid rise in sea levels, difficulty growing enough food, huge die-offs of forests, and mass extinctions of plant and animal species."[10] The report goes on to warn that only with an intense

[9] *Climate Change: Evidence & Causes* (2014), p. 15. [10] Gillis (2014).

push over the next fifteen years can we manage to keep planetary warming to a tolerable level. Ottmar Edenhofer, co-chair of the committee that wrote the volume of the report on the impacts of climate change, stated in April 2014, "We cannot afford to lose another decade. If we lose another decade, it becomes extremely costly to achieve climate stabilization."[11]

12.3 The Prisoner's Dilemma and the Tragedy of the Commons

So far, after several decades of increased understanding about the causes and effects of climate change, the global community has failed to take any meaningful action in response. Why is this? In advanced industrialized countries and in large developing countries like India and China, burning fossil fuels is deeply woven into the fabric of our lives. Greenhouse gases are emitted from hundreds of millions of individual sources, as we drive our cars, fly our planes, heat our homes, and use electricity from coal-fired generating plants to light our buildings and run our machines. The emissions from these millions upon millions of sources intermingle and become indistinguishable in the atmosphere. They produce climate change, most of the effects of which happen much later and in locations all over the world. Nothing that any individual or firm can do alone will have any appreciable effect on the problem. And with only some exaggeration, nothing any single nation does will solve the problems either. The problem of climate change is global and intergenerational, and the solution must be global and intergenerational as well.

The global problem has the structure of a Prisoner's Dilemma or, more precisely, a variation of the Prisoner's Dilemma known as the Tragedy of the Commons. The Dilemma involves a conflict between individual and collective rationality. In one version of the classic Prisoner's Dilemma situation, two individuals are arrested on suspicion of having committed several crimes together, and each of them is given the option of confessing or not confessing. Each prisoner has an overriding interest in minimizing his sentence if convicted, and for each prisoner the chance of being convicted of the more serious crime, and the sentence he receives if convicted of either crime, will be determined by whether he confesses and gives testimony that can be used against his confederate, and whether his confederate confesses and gives testimony against him. Each prisoner faces the identical situation.

[11] Gillis (2014).

It turns out that each prisoner will receive a lighter sentence if he confesses, regardless of what his confederate does. It is therefore rational for each prisoner to confess, but if both confess, they will each be found guilty of the more serious crime, and the sentence for both of them will be worse than the sentence each would receive if neither confessed, because without the testimony of a confessor, each could be convicted only of the lesser crime. That's the Dilemma.

Hobbes' political theory has the structure of the Prisoner's Dilemma. In the state of nature, it is in the interest of each individual to keep and exercise all of his natural rights, whatever other individuals do. If no one else surrenders his rights, you would be foolish to do so; and if everyone else surrenders his rights, you can gain a further advantage by refusing to surrender yours. The result is that each individual's pursuit of his own rational interests means that each keeps the right to do anything to anyone. No cooperation is assured, and everyone lives in a state of insecurity that Hobbes famously describes as "a war of all against all." The life of each man in this state is "solitary, poor, nasty, brutish, and short."[12]

There are a number of possible ways to solve or avoid the Prisoner's Dilemma. If the goal is to get everyone to agree on a solution that is both collectively best and also better for each individual than the outcome that results from each agent choosing what is individually best regardless of what the others do, then one needs to find a way to punish defection from the collectively best choice or otherwise create incentives for cooperation that are stronger than the incentive to defect. If defection can be punished, then the payoffs can be changed and the Dilemma avoided. If there is a natural basis for trust among the players that is more strongly motivating than the desire to maximize one's individual gains, as there may be within a family or a set of friends or allies, then individuals may find it reasonable to choose the collectively best alternative, hoping that the others will do the same and not defect. If the situation (or "game") is going to be repeated often, then the rational choice for each player may be to communicate a willingness to cooperate by cooperating in the first game, and then, if others cooperate, continue to cooperate in repeated games so long as the other players also continue to cooperate. A convention of cooperation may thus be established that each player realizes is in his best interests to support, a convention that becomes more secure with repeated instances of cooperation.

[12] Hobbes (1651), ch. 13.

Finally, if the preference for reaching agreement is itself stronger than the preference of each individual for his most preferred outcome, then the game has changed, the Dilemma is avoided, and the parties should be able to agree on the collectively best solution. For example, if two friends living in Chicago want to go to a baseball game together, but one prefers to see the Cubs play while the other prefers to see the White Sox, they may each nevertheless also prefer to go to a baseball game together than to see their favorite team play. They should then be able to work out a cooperative solution, perhaps agreeing to see the Cubs play today and the White Sox next time.[13]

None of these solutions is available in Hobbes' state of nature, however, so he argues that each person has a rational obligation to surrender all his rights and powers to a Leviathan, on condition that others similarly surrender their rights. The Leviathan then accepts the responsibility for bringing the individuals out of the state of nature by forming civil society and protecting them from their natural fate.

Prisoner's Dilemmas can be solved or avoided, then, if there is a strong natural basis for trust among the players, if they can be punished for defection, if they are motivated by moral ideals that are stronger than their reasons to pursue their own self-interest, if they all know that they will have to continue doing business with each other and should thus establish a convention of cooperation, or if the players prefer cooperation itself to the other possible outcomes. Otherwise, solving the Prisoner's Dilemma may require surrendering one's rights and sovereignty to an agent like a Leviathan that can compel everyone to cooperate for the good of all rather than allowing each to pursue his individual best strategy, which leads to an outcome that is worse for each of them.

The Tragedy of the Commons is a variation of this problem. In the classic example, herdsmen graze livestock on the commons. Once the commons has reached its carrying capacity for grazing, any additional stock will reduce the availability of food for all the animals and thus lower the value of the entire herd. But each herdsman nevertheless has a rational incentive to add another animal, because he will capture all the benefits of overgrazing, while the costs will be distributed among all the herdsmen. The average value of each animal in his herd will decrease slightly, but the value of adding another animal will more than offset the losses that he alone will suffer. All the herdsmen have identical incentives, so each has a reason to add an animal to his herd,

[13] This situation, under the name "the Battle of the Sexes," has been well studied in the game theory literature.

and then another, and so on, until the system crashes or the quality of all the stock diminishes to the point that all the herdsmen are worse off than they were before the first animal was added that exceeded the carrying capacity of the commons.

The possible solutions to the commons problem are analogous to those of the Prisoner's Dilemma. One can try to change the structure of the problem by eliminating the commons, either by privatizing it, so that the externalized costs of overgrazing could be taken into account, or by adopting some authority – a Leviathan – to regulate and enforce grazing policies. These two solutions were what Garrett Hardin, who famously described and gave the name to the problem, proposed. [14]

One significant difference between the Prisoner's Dilemma and the commons problem is that the former consists of a single binary choice (cooperate or don't cooperate – confess or don't confess), or perhaps repeated games each consisting of a single binary choice, while the latter consists of incremental decisions. Each herdsman must decide whether to add an animal, then another, then another, and so on. For this reason the commons problem more closely describes the climate change situation. Each actor – individual, firm, or nation – decides to increase its emissions in order to attain greater consumer or economic benefits, and then each actor increases them again, and so on. For example, as China pursues its policies of economic development, it is currently adding about one coal-fired electric power plant per month. The result of each increase is a net benefit to the actor, because she receives most of the benefits while the costs are externalized, in this case to everyone on earth, now and across generations.

If other nations were to agree on a policy that required them to reduce their emissions, in part by shutting down their coal fired electric generators and replacing them with other, cleaner but more expensive sources of electricity, China would have some incentive to cooperate, but it would have perhaps an even greater incentive to continue building cheaper coal-fired plants. China might further claim that it has a moral justification for its non-cooperation: it is only using the means that other, wealthier nations have used for several centuries to develop their own economies. It is unfair, the Chinese might claim, for the developed nations to try to inhibit China's development in order to solve a problem that was created primarily by the developed nations. If just China and India, with their large populations, decided that their

[14] Hardin (1968).

reasons for pursuing economic development were stronger than their reasons for cooperating with other nations in reducing greenhouse gas emissions, they could effectively negate the intended consequences of policies the developed nations could feasibly agree on to reduce emissions to a level that scientists believe we could reasonably manage. Moreover, if China and India refuse to cooperate, then the incentives are weakened for other nations to accept the costs of cooperating instead of continuing to compete to achieve their own greater rates of economic growth.

As this dynamic continues, the situation becomes worse for everyone. We all have a rational incentive to reach an agreement to limit emissions, but without strong bonds of trust or means for enforcing the agreement and punishing those who would defect in order to further increase their own GDP, or unless individual nations could be motivated by considerations of justice or other moral reasons to accept the costs of limiting emissions and mitigating the inevitable harms of past and current emissions, each nation has a strong reason to defect from (or weaken the terms of) any proposed agreement that might solve the problem. And the situation continues to worsen.

It is worth noting that Hardin insisted that there is "no technical solution" to the Tragedy of the Commons. He meant that there is no possible technological fix that would allow the commons to remain a commons and yet would not lead to exceeding its carrying capacity. When Hardin published his article in 1968, his main concern was overpopulation. People around the world were more or less free to make their own reproduction decisions, and the earth – its water and its resources for producing food – was like a commons that Hardin thought would soon exceed its carrying capacity as people, especially in the poorest countries, continued to reproduce without limit. He claimed that his analysis showed that we needed some strong policies for limiting population growth, especially in poor countries where birth rates are the highest, in order to fend off a pending global disaster.

Hardin turned out to be wrong in his Malthusian population projections. For one thing, he did not realize how incentives to have more children worked. He did not consider, as most demographers and historians now insist, that poverty causes population growth, rather than the other way around, and that economic development could help bring down birth rates. He also misunderstood the nature of the carrying capacity of the earth and failed to appreciate how much technological development can change what the carrying capacity is. Many economists now insist, for example, that reserves of non-renewable resources are effectively infinite because of technological progress.[15] The limits on the atmosphere to absorb greenhouse gases without causing harmful climate change may be an important exception to the optimistic view that the resource base is for practical purposes unlimited, but Hardin did not consider this issue. The point is that he never explained why technological developments, including genetic engineering, could not be applied to change the nature of the vegetation on the commons or the nature of the livestock who live on it in ways that could drastically expand the "natural" carrying capacity.

Hardin's view was pessimistic, almost apocalyptic, as are the views of some environmentalists today who insist that climate change requires us radically to change the way we live and are appalled by the prospects of proposed technological or geoengineering solutions to the climate change problem.[16] I believe that the natural incentives and moral reasons that apply to climate change are not likely to move us to solve the problem until some of the worst effects are within sight and beginning to occur. At that point we – or our heirs – will need every means we can get our hands on to ward off the possibly catastrophic consequences. If this view is correct, then we should be advocating changes in our behavior at the same time as we try to develop technological and engineering solutions to some of the problems we otherwise fail to prevent.

12.4 Intergenerational morality and discount rates

Many actions have environmental impacts that become noticeable only further in the future. For example, the real damage to ecosystems from our practice of killing predators became evident only much later. What makes human induced climate change unique is not the fact that greenhouse gas emissions have long-term consequences; rather, it is the scale of the consequences and our increasing ability to predict them now.

Game theory can explain the global aspects of anthropogenic climate change, because game theory applies to contexts where the results of one's choices are determined in part by the choices others are making. A kind of reciprocity is built into the structure of games, as well as the need for coordination and communication. But for obvious reasons, this kind of analysis does not help us in reasoning about the intergenerational dimensions of the problem. Our actions will affect people who live in the future,

[15] See Sagoff (2008). [16] See, for example, Clive Hamilton (2013); also Gardiner (2011).

but beyond the lives of our children and grandchildren, there is virtually no reciprocity in our relationship with them.

There is one important exception to this fact, one way in which backward causation is possible in normative matters or the world of values. Our heirs have control over our reputations. They also have control over the fate of projects that we value which give meaning to our lives but which will not be completed in our lifetime. These include projects aimed not at increasing our happiness or welfare but at doing something good in the world. Although we may know that some of these projects will not be completed in our lifetime, it matters to us that we expect and hope that they will be completed in the future. The cathedral will be built; the cure for the disease that we are studying and trying better to understand will someday be found, and so on. Our heirs also have control over the survival of cultures, languages, and important traditions. How they will exercise their powers over us of course depends on their attitudes toward what we have done and what we have left them, as well as the resources they will have to act on the reasons these attitudes support.

When philosophers, economists, and policy makers talk about the relation of those who are alive today to people who will live in the future, they often speak as a kind of shorthand of relations between generations: how the current generation's actions will affect the next and succeeding generations, what we owe future generations, and so on. This simplification is perhaps unavoidable for building theories and models, but it can also be misleading. Generations don't act, and they are not the kinds of entities that can care or respond to reasons or have moral obligations. When we say that Generation X has certain concerns or values, we are speaking elliptically. We mean that a higher percentage of individuals in that generation have those concerns or values than their predecessors. Only individuals and organized groups such as firms or nations can be agents in this sense. The people who live in the future will not be organized into generations that act as a unit but into families, cultures, and perhaps nations, as we are today.

This fact explains why it is highly misleading and perhaps nonsensical to talk about intergenerational justice. That makes no more sense than it does to speak of environmental justice, if what is meant by the latter is not just the obvious fact that environmental policies can be harmful and unfair to certain groups of people, but the more encompassing idea that duties of justice should govern our relationship with plants, trees, and ecosystems.

Without the possibility of reciprocity, the foundation for any reasonable theory of justice is lacking. This means that we will have to approach our reasons for caring about what happens in the distant future in some other way.

Two facts make the philosophical problem of intertemporal rational or moral choice particularly difficult. The first is the problem that economists are most aware of, which is that the future lasts a long time and will include many, many people. This fact creates some particularly thorny issues for doing cost–benefit analyses to determine which large investments today are economically efficient. If an investment today were to produce benefits that extend indefinitely into the future, no matter how small those benefits are, then their sum would eventually outweigh any cost today, no matter how high. If we simply added costs and benefits across time to justify our actions, it would have the effect that people living today would be slaves of the future. To avoid this consequence, economists insist on discounting costs and benefits as they occur further in the future. But what are the justifications for discounting? And what is the correct discount rate?

Economists claim that reasons of economic efficiency support discounting, and nobody disputes these arguments. When an actor makes an investment – whether the actor is an individual, a firm, or a government – it is efficient to receive returns on investments earlier and defer costs to later. The value of $100 ten years from now is less than the value of $100 today. This is because as the economy grows, the price of commodities falls. Thus, there are opportunity costs to consuming now rather than investing now and consuming later, and these opportunity costs are measured by a discount rate.

Some philosophers also defend a different, moral reason for discounting. Most economic analyses predict that even with climate change, the world's economy is likely to continue to grow in the future, although at a lower rate than it would grow without climate change. There is some possibility that climate change will produce disasters that will reverse economic growth worldwide, but economists who have looked closely at the scientific evidence think this possibility is still unlikely (though not to be ignored). If the economy continues to grow in real terms, then people in the future will be wealthier than we are today. This means that paying costs today to compensate for the harms of climate change involves transferring resources from people who are poorer to people who are wealthier. As a matter of justice, some philosophers argue, we should count the marginal benefits and costs to people who are better off for less than the same marginal benefits and costs to people who are worse off, in order to give priority in our policies to those who are worse off. This view of distributive justice is known as prioritarianism.[17]

But even if we accept these and perhaps other reasons for discounting, the question of the proper rate remains. An individual or firm, whose only goal is to maximize profits, may find it rational to discount at the market rate of return on capital, which turns out to be in the neighborhood of 5–6 percent annually. Should this be the discount rate used for public investments as well, such as those necessary to reduce or mitigate the effects of climate change? The answer depends on two things: the proper goals of public investment, and what determines the market rate of return.

The rate of return is determined by the cost of borrowing money, which is a function not only of opportunity costs but also individual time preferences. If most people simply prefer to consume things earlier than later, then this preference will partially determine interest rates. Your preference for buying a new car or a flat screen television now rather than later is expressed by what you are willing to pay to borrow money to make these purchases today.

We should insist that our governments spend money efficiently, but this does not mean that government investments should be based on the same reasons that govern profit-maximizing firms or individuals. We expect our government's policies also to express certain moral principles. Government actions are necessary to maintain national security, protect individual rights, meet the demands of justice, protect the environment, and preserve areas of natural beauty and cultural or historical significance. These are things we value as citizens, not as consumers, and cannot satisfy individually. However, we also expect our government representatives to express the will of the people they serve or to respect citizens' sovereignty in proposing and voting on laws and regulations. And, as I have just described, most people have time preferences: they prefer to consume or receive benefits sooner and defer costs to later.

Expressing values that we cannot express as counsumers, and respecting citizens' sovereignty are different goals that make different demands and can generate conflicts about the proper role of government. One instance of this conflict was dramatically on display in a debate over the economics of climate change. In 2006, a committee commissioned by the British government and

[17] For excellent discussions of discounting and prioritarianism in the context of climate change, see Broome (2012).

chaired by Sir Nicholas Stern to study the economics of climate change issued a report, which concluded that economically developed countries like those in Europe and the US should be making some costly investments now to reduce carbon emissions and mitigate the future effects of climate change.[18] The report argued that the benefits of these investments outweighed the costs. Some other prominent economists, mostly in the US, took issue with the Stern Report. These critics argued that, although governments should certainly act now to put a price on carbon so that the social cost of emitting greenhouse gases would be internalized and taken into account in private investment and consumption decisions, it was not economically reasonable at this time to make more costly public investments to mitigate the future effects of climate change.[19] The costs of these investments outweigh the benefits. The critics recommended that, apart from putting a price on carbon, we should continue with "business-as-usual," including investing in scientific research and technological development that will make it more efficient to respond to climate change in the future.

The analytic assumption that divided Stern and his critics, as both sides acknowledged, was the discount rate. Stern claimed that reasons for discounting for public investments should be based on rational and moral arguments, and he agreed with some philosophers and economists who have argued that pure time preference is irrational, that it reflects imprudence and can be used as an excuse to cheat the future. Therefore, pure time preference should not be a factor in determining the discount rate. Considering only reasons of efficiency, principles of justice like prioritarianism, and making a small allowance for uncertainties, Stern concluded that the proper discount rate for government investments was approximately 2 percent. The critics argued that government officials and economic advisers should not be substituting their own value judgments for the expressed preferences of the public, so the discount rate should be determined by the market or the rate of return on capital, which is approximately 5 or 6 percent. This difference dominated all other considerations in the respective analyses, and it fully explains the disagreement between Stern and his critics.

[18] Stern (2007).

[19] See Nordhaus (2007) and Weitzman (2007). It is important to add that both Nordhaus and Weitzman have become increasingly strong advocates of the economic arguments for taking strong actions now, such as imposing a tax on carbon. Their more recent arguments reflect further developments of climate science. See, for example, Nordhaus (2013); and Weitzman (2013).

The philosophical issue at the heart of this debate is an important one about the nature of democracy. When we say that democratic laws and policies should reflect the will of the people, do we mean that they should reflect our preferences, such as those we reveal in our economic behavior or opinion surveys? Or do we mean that politicians and policy makers represent us best by making decisions that they judge to be best for the nation? In the latter case, we are asking our politicians and policy makers to weigh reasons for their decisions and justify them with arguments. We would be asking our representatives to act in ways similar to judges, who write opinions to explain and justify their decisions.

Although some political theorists disagree, it seems clear to me that democratic governments ought to do more than serve interest groups or reflect popular or partisan opinions. The foundation of democratic governments, after all, as reflected in the various constitutions of democratic states, rests on philosophical and moral principles. These include the constitutional protection of basic individual rights, which are meant in part to shield individual freedoms from laws or policies that may reflect the preferences of the majority but would unduly restrict individual rights. They also include the justification for programs like social security and other measures of the welfare state that provide a safety net to protect individuals not only from the unfair consequences of fate and bad luck but also from the consequences of our own impatience, such as the time preferences that lead most young people to spend more and save less for their old age than rationality and prudence would warrant. Finally, on this conception of democracy, discount rates for public investments should be determined by justifiable reasons for discounting, not by consumers' untutored and often unreflective time preferences.

12.5 The non-identity problem

I will return in the final section of this chapter to discuss some implications for climate policies of this debate about the nature of democracy, but I turn now to a second fact about intertemporal moral reasoning. This is the "non-identity problem," which has been discussed and argued at length by philosophers.[20]

When an individual knowingly or needlessly acts in a way that harms another person, justice requires holding him morally responsible. He should

[20] See Parfit (1984). For discussion in the context of climate change, see Broome (2012).

compensate the victim, be punished, or both. It is irrelevant from a moral point of view how close or far away the victim is, either spatially or temporally. But when a powerful nation or the international community acts or fails to act, the situation is more complicated. Actions on a large scale or actions with powerful effects (including possible actions by individuals, such as Hitler) can produce consequences that ripple and expand with time. Such actions affect future populations in a way that determines the identities of the people who will live. Had it not been for World War II and the effect of that war on the people alive then, for example, my parents would not have met or conceived children when they did; neither my wife nor I would have come into existence; and nor would any of our heirs. In this sense, World War II has determined the identities of most of the people alive in the world today. Actions on a sufficiently large scale thus determine both the identities of the people who will live in the future and their circumstances. Plausible principles of justice or conceptions of harm, however, are based on a person-affecting principle. This means that people cannot be harmed or treated unjustly unless they are treated in ways that make them worse off than *they* would otherwise have been.

Climate change decisions – for example, whether the international community takes significant action soon to curb global emissions or continues with business-as-usual – have this kind of effect. This means that if we take no action today, and people who live a century or more in the future suffer as a result, they cannot claim that we have treated them unjustly or unfairly. Had we acted differently, they would in all likelihood not have existed. A different population of people would instead have existed under different circumstances. This non-identity effect is one reason for casting doubt on the coherence of claims about intergenerational justice with respect to climate change. Of course, this does not mean that we have no moral responsibilities in these situations or that what we do is not wrong if the people we cause to exist live in horrible circumstances that are the consequence of our failure to act. But our moral reasons and responsibilities in these situations have to be explained in some different way.

In his book on the ethics of climate change, John Broome argues that the requirements of public morality and personal morality with respect to climate change divide rather sharply: governments have duties of goodness but not justice, and individuals have duties of justice but not goodness.[21] The reason

[21] Broome (2012).

governments do not have duties of justice, according to Broome, is because of the non-identity effect. But governments generally have strong duties to prevent bad effects and to improve the wellbeing of their citizens. When a government builds an airport, a highway, or a school, for example, it does not compensate all the individuals that will be harmed by these actions. It will typically compensate those whose property is condemned and thus suffer the greatest harms, but it will not compensate those who suffer from more noise, traffic, etc. This is because governments, according to Broome, have strong duties to promote the good or the welfare of their citizens that permit them to act in certain ways that would otherwise be unjust.

Individuals, in contrast, do not have duties of goodness with respect to climate change, because there is little any individual can do to make things better or to prevent the serious damage of climate change. But Broome argues that individuals do have duties of justice with respect to our greenhouse gas emissions. This is because the lifetime emissions of each person living today contribute to the overall increase in greenhouse gas concentrations. Our individual contributions, however, are not significant enough by themselves to produce the non-identity effect. So each of us contributes to the harm of climate change, and the effects of our individual actions will increase, even if only by a minuscule amount, the bad effects that each of billions of people in the future will endure as a result. Furthermore, Broome argues, anyone living today in an economically developed country could feasibly reduce her individual carbon footprint to zero without serious sacrifices to her quality of life, by a combination of reducing some emissions and offsetting the rest. Broome argues that we each therefore have a duty of justice to reduce our net emissions to zero.

One might doubt the feasibility of offsetting all of our individual emissions if enough people in the developed world took offsetting seriously as a moral responsibility. But a more serious objection to Broome's argument, I believe, is whether we have duties of justice to avoid harming individuals in the minuscule way that any person's emissions today will contribute to the harm suffered by any individual in the future. What plausible theory of justice holds us to standards this strict or demanding? Minuscule harms are often regarded not as unjust acts but as slight inconveniences, which may warrant an acknowledgment and an apology, but not compensation, punishment, or the other demands of justice. Absent a plausible account of justice to underwrite this conclusion, which Broome does not give, we are left to wonder about the nature of our moral responsibility or moral reasons in this area.

12.6 Political failure

The global and temporal dimensions of rational and moral choice help to explain the difficulties of finding practical solutions to the climate change problem. They help to explain our failure in this area to put long-term global needs ahead of short-term local interests. Even as the cause and likely effects of climate change are becoming clearer to us, we have failed so far to respond in a meaningful way.

Focusing on the US alone, the *New York Times* reports that, "Polls consistently show that while a majority of Americans accept that climate change is real, addressing it ranks at the bottom of voters' priorities."[22] This is because Americans are more concerned with unemployment, the rising cost of health care, fighting terrorism, and other problems that are closer to home. And these political preferences are reflected at the national level. Leaders in the fossil fuel industries, moreover, have enormous political power, and in their pursuit of furthering their own short-term goals, they spend heavily to punish politicians who support policies that would hurt their profits. Thus, while the effects of climate change are being felt everywhere in the US, the Congress has allowed tax credits to encourage renewable sources of energy to lapse and continues to give tax breaks to the oil and coal industries. According to the International Monetary Fund, the US government is the world's largest single source of subsidies to the fossil fuel industry worldwide, which accounts for an estimated 13 percent of carbon emissions.[23]

Some states and regions in the US that have recently experienced some of the worst floods, snow storms, or droughts in their history have begun taking action by setting emission quotas and establishing cap-and-trade systems to give industries in their states an incentive to reduce their own emissions. And President Obama is attempting to use his executive power to reduce emissions from automobiles and inhibit the building of more coal-fired electric generating plants. These measures make some Americans more hopeful that the US is finally prepared to lead the way in responding to climate change. But most experts insist that these measures by themselves are an inadequate response, and that nothing will significantly reduce emissions in the US without a nationwide carbon pricing policy. As Michael Bloomberg, former mayor of New York City, said in a speech supporting a carbon tax, "If you want less of something, every economist will tell you to do the same thing: make it more

[22] Davenport (2014). [23] See International Monetary Fund (2013).

expensive."[24] But the political prospects of that happening in the United States in the foreseeable future are negligible.[25] Moreover, without the United States taking effective action to raise the price of carbon, it is more difficult to convince other nations to enact and follow through on policies to reduce their emissions.

At the international level, the rhetoric supporting climate change agreements has been more encouraging, but the results so far have not. In response to the developments in climate science as reported in the First Assessment Report of the IPCC, the nations of the world sent representatives to the Rio Earth Summit in 1992 and signed the United Nations Framework Convention on Climate Change (UNFCCC).[26] This Convention called for stabilization of greenhouse gases in the atmosphere that would "prevent dangerous anthropogenic interference with the climate system," and it called for "common but differentiated responsibilities" that would require developed countries to take the lead in reducing emissions while developing countries would continue developing and take significant action later. Another set of meetings aimed at making the goals of the UNFCCC more specific led to the 1997 Kyoto Protocol, which called on industrialized nations to reduce their emissions to below their 1990 levels. The US signed the Protocol, but the US Senate refused to ratify it. Meanwhile, continued negotiations and compromises were needed to get the required number of signatures from other states to give the Protocol the force of international law. But the agreement began to fall apart in these negotiations, and no enforcement mechanisms were built into it. The result is that, except for the European Union, aided by an economic recession that cut economic production and thus reduced emissions in Germany and England, none of the participating nations succeeded in meeting its goal. After twelve years of failing, another summit met in Copenhagen in 2009. It turned out to be an embarrassing failure, and no new agreement was approved or signed.

In the meantime, even as technologies that will create more energy efficiency and allow a shift to renewable sources of energy are quickly becoming more affordable and practical on a large scale, global emissions of greenhouse gases are increasing at a faster rate than ever. Atmospheric CO_2 levels in the

[24] Quoted in Kolbert (2014), p. 21.

[25] See Kolbert, ibid. For an interesting exchange on the politics of a carbon tax, see Paulson (2014); and Krugman (2014).

[26] For a summary of international meetings and agreements on climate change, see Gardiner (2011), Part B.

first decade of this century rose at nearly twice the rate of the last decade of the twentieth century, and while rich countries are making slow progress in cutting their emissions, developing countries like China and India are rushing to build new coal-fired power plants.[27]

12.7 The tension between global and intergenerational solutions

The current international stalemate is an illustration of the Prisoner's Dilemma or the commons problem. It is not that individual actors (in this case primarily nations) are morally corrupt, as the executives of some fossil fuel companies may be. Rather, it is that in the absence of a mechanism of international enforcement, domestic political pressures, which are more self-interested, continue to triumph in most countries over policies that would require some national sacrifice for the sake of the greater global good. And, as in the commons problem, the long-term outcome is worse for each nation.

This suggests that a solution to the global problem might require something like a Hobbesian Leviathan, a world agency or government that could allocate responsibilities and enforce agreements that would force individual nations to put global health ahead of their own narrow interests. In other words, the solution appears to require something that can force us to respond to the problems as a global community and make us see ourselves as global actors, citizens of the earth, not just members of our own local tribes. Framed the right way, this alternative can sound like a noble calling and, who knows, it may even garner some political support.

But this proposed solution to the global problem, even if it were feasible, is undermined by the temporal problem. Intergenerational moral reasons, as I have explained, are marked by a lack of reciprocity: What can the future do for us? Past generations have caused many of the problems, which they have passed on to us. We can either act in ways that will begin to solve them, or "pass the buck" to our heirs, who will face bigger problems. The intertemporal question is: How can we effectively be motivated to restrain the pursuit of our near-term interests for the sake of our descendants?

As I suggested above, I believe a natural motivation to care about one's descendants is rational, widespread, and strong. It springs from the kinds of values that attach us to the world and to the projects to which we contribute but which we do not expect to be completed or to reap benefits in our

[27] Gillis (2014).

lifetimes. These are the kinds of projects we value because they allow us to live more meaningful lives. Parents often willingly sacrifice for the sake of their children, their grandchildren, and their further descendants. Individuals commonly see their own societies, in Edmund Burke's words, as "a partnership not only between those who are living, but between those who are living, those who are dead and those who are to be born."[28] We each have our selfish and narrow interests, but most people also see themselves as links in a chain that connects past, present, and future. We inherited much of what we value from our ancestors; we modify and add to these valued things; and we pass them on to our descendants, hoping that they will respect, treasure, and continue to improve on what we have found valuable. This kind of attitude explains our concerns to support basic research, to protect and preserve historical and cultural artifacts, the natural environment, and the traditions that we find noble. We value our identity with these things in part as a way of connecting us to something beyond our own experiences and allowing us to make a positive mark on the world that may continue to be felt and influence human development after we die.

In a recent book that brilliantly explores these issues, Samuel Scheffler argues that our lives would lose much of their meaning if we became aware that human life would end shortly after we died.[29] Although future people cannot enter into bargains with us, they hold important cards that can naturally motivate us to take their interests into account.

If this motivation to care about the future is natural and implicitly alive in us, as I am claiming, how can it be made salient, so that we can care enough to restrain some of our narrower consumer interests for the sake of leaving our descendants a better world? Granted, our motivational springs are often more responsive to proximate desires than they are to more enduring values. We want the new car today, not tomorrow when it will cost less. But the evidence that other values are also salient is the fact that people and groups commonly make sacrifices for their descendants. A complicating factor about these deeper motivations, however, is that we naturally care more for our families, our descendants, and our fellow tribe members, than we care about humanity at large. Our deeper values are often tied to particular cultures and traditions, not to our recognition of ourselves as citizens of the world.

If this is right, then there appears to be a tension between the most plausible solution of the global problem and the most feasible solution of

[28] Burke (1790). [29] Scheffler (2013). I have made a similar argument in MacLean (1983).

the temporal problem.[30] The global problem demands a solution that requires countries like the US, which have contributed most to causing the climate change problem, to contribute most to solving it. We will have to cut our emissions and take the lead in inducing other industrialized countries to cut theirs. We will also have to take the lead in mitigating the effects that we cannot prevent, including helping the poorest and most helpless people and countries to respond and adapt to the worst effects of climate change. And we will have to help developing countries like China and India to achieve their goals of modernization without relying on emitting greenhouse gases as we and all the other economically advanced countries did in the process of our industrialization and economic development. But the temporal problem requires us to be more aware of our particular cultures and traditions and why we value protecting them for the future.

12.8 Conclusion

Burke believed that if the present generation were to act as a trustee of the interests of the past and the future, it would have to abandon democracy for rule by a natural aristocracy, which could mediate conflicts among generational interests by choosing what is best to conserve and passing it on. Other commentators on the climate change problem have also suggested that solving it may require severely restricting our democratic powers.[31] I am suggesting that an alternative to abandoning democracy is to modify our understanding of what democracy is. If our elected representatives could free themselves from being captives of the narrow economic interests of those who are most wealthy and powerful, and if our representatives would promote the idea that the meaning of democracy implies that elected leaders should protect and help articulate the nation's deepest values, rather than catering exclusively to narrow and partisan pressures, the US and other developed democratic countries could begin addressing the temporal problem of climate change and assume leadership in solving the global problem without surrendering sovereignty to a world government.

I have described in outline what I think is a possible solution to the two main philosophical challenges of solving the climate change problem, but I am not naive enough to expect that a practical solution will be easy or likely

[30] This tension has also been provocatively explored in a review of Gardiner (2011). See Bull (2012).

[31] See Bull (2012).

to come about in the near future. There are four possible options for responding to climate change. One is to change the pattern of energy production around the world in order to limit and then reduce overall emissions, which will not avoid all the harms that our emissions will inevitably produce but could avoid the worst disasters; a second option is to do nothing, continue with business-as-usual, and simply accept what comes; a third is to focus especially on attempting to mitigate the changes that will occur and figuring out how best to adapt to them; and the fourth is to try to find technological fixes or "geoengineering" solutions to counteract some of the changes that would otherwise occur. The joint report of the Royal Society and the US National Academy of Sciences lists these options and concludes that "each has risks, attractions, and costs, and what is actually done may be a mixture" of them.[32] Although the language of this report is genuine committee-speak, this conclusion is good common sense. Given the unique temporal lag between cause and effect in climate change, it may well turn out that we will be unable to find the motivation actually to take strong enough actions to avoid the consequences and mitigate and adapt to their effects until some of the worst damages are at our door. The latest scientific findings suggest that some of these damages may begin happening sooner rather than later. In any event, it is likely that we will need all the tools we can lay our hands on. If this scenario is at all likely, then scientists and engineers will inevitably have a very large and important role to play. We should be supporting them even as we try to change our own behavior and our political culture.

[32] See *Climate Change: Evidence & Causes* (2014), p. B9.

13 Prisoner's Dilemma experiments

Charles Holt, Cathleen Johnson, and David Schmidtz

For decades, social scientists have been studying empirical tendencies of human subjects to cooperate in laboratory Prisoner's Dilemma settings. Here is what we are learning.

13.1 Introduction

In a classic Prisoner's Dilemma, players choose independently whether to contribute toward a mutually desirable outcome. Cooperation comes at a cost, so that (for example) a dollar's contribution yields a return of more than a dollar to the group but less than a dollar to the individual contributor. Thus contributing is optimal for the group and suboptimal for the individual. Not contributing is a *dominant strategy*, which is to say, standard game theory predicts defection. If players prefer higher to lower monetary playoffs, and if other things are equal so far as a player is concerned, a player will not contribute.

And yet, among the more robust findings to come out of experimental economics is that human players do not behave as game theory predicts. In laboratory Prisoner's Dilemmas, individual contributions toward the group good can be substantial.[1]

We thank Taylor Apodaca, Emily Snow, Sam Brock, Flor Guerra, Tori Morris, Jacob Scott, and Shelby Thompson for helpful comments and research assistance. This research was funded in part by the Earhart Foundation, the Property & Environment Research Center, and the Templeton Foundation. We also thank our editor, Martin Peterson, for his patience and helpful suggestions.

[1] Note that we define the Prisoner's Dilemma in terms of a matrix of plainly observable (that is, monetary) payoffs rather than in terms of preferences. The cost of defining the Prisoner's Dilemma as a preference ranking (as many theorists do) is that whether an experimental design counts as a Prisoner's Dilemma would depend on what subjects are like (how they choose to rank outcomes) rather than on what the payoffs are like. By contrast, defining the Prisoner's Dilemma in terms of controllable and determinable (typically monetary) payoffs

In iterated Prisoner's Dilemmas – multi-period games where each period viewed in isolation has the structure of a Prisoner's Dilemma – game theory does not so straightforwardly predict defection. Iterations allow players to implement strategies of reciprocity, including the simple but effective strategy of *Tit-for-Tat* (Axelrod 1984), which means responding to cooperation by cooperating and responding to defection by defecting. In iterated laboratory games, the phenomenon of human subjects making substantial contributions is likewise robust. However, it also is a robust result that cooperative relations tend to crumble in an iterated game's final period (when players know in advance that a given period is the final one). Typically, levels of cooperation remain positive, but decay in the direction predicted by standard game theory.

Laboratory experiments can test *theories* about how people behave in a Prisoner's Dilemma. Experiments also can test *policy proposals*. When testing a theory, we design an experiment so that a target theory yields testable predictions about what we will observe in the experimental setting. If predictions are disconfirmed, then our theory needs rethinking. When testing a policy, we design an experiment not to test a theory but to replicate a policy environment, and then check to see how a proposed policy works in that simulated policy environment. If we find that a policy (designed to encourage cooperation, for example) instead creates unanticipated incentives or opportunities that result in something altogether contrary to the intent of the policy proposal, then the policy needs rethinking (Plott 1982).

As a test of theory, laboratory results are intriguing. Human players do not act as predicted by standard game theory. It can be hard to say why. Perhaps subjects do not understand what, to laboratory personnel, seem like simple directions. Perhaps some subjects do not think in the strategic way posited by game theory but instead make decisions by heuristic. Instead of calculating their best strategy a priori, perhaps they too are experimenting; that is, they divide their money into several pots and then see what happens. Finally, subjects may have views about the morality of what they are doing. Concerns about morality can enter an economic calculation as goals or as constraints. Players may care enough about fellow subjects to consider the good that their contributions does for others. Alternatively, morality as a constraint may lead players to view being a free rider as out of bounds even if they otherwise aim to make as much money as they can. (On the downside, some subjects may

enables social scientists to treat whether subjects will cooperate in a Prisoner's Dilemma as a straightforwardly testable empirical question.

play to win rather than to make money, and may attach positive value to *reducing* their partners' payoffs.) In any case, what motivates human agents in laboratory Prisoner's Dilemma settings seems complicated indeed.[2]

Extrapolating, we know that standard game theory's prediction of end-period defection often is near the mark when partners are utterly anonymous. We also know from the larger laboratory of life experience that people who rely on networks of stable cooperation do not simply accept conditions of anonymity but instead gather information about prospective partners. People work to create conditions where even the final period of a given partnership can in some way still cast a shadow over a player's future. Suppose final period cooperators end up with (something analogous to) a high credit rating whereas final period defectors end up with a low credit rating. In that case, final period cooperation enhances one's chances of attracting high-quality future partners. It stands to reason that such information makes for a more cooperative society.

Accordingly, we gossip. Or, where an institution such as the internet makes relationships so distant and anonymous that there is no opportunity to gossip face-to-face, other institutions emerge to provide customers with opportunities to file complaints, to provide ratings of customer satisfaction, to provide lenders with credit scores, and so on. In short, we work to ensure that a given decision can affect a person's reputation.

We also know that people, trying to protect themselves from free riders, develop institutions in which decisions are made jointly rather than independently. Buyers do not want to pay and then be left simply hoping that sellers will deliver. Sellers do not want to deliver and then be left simply hoping that buyers will pay. So, buyers and sellers develop institutions in which their moves are contingent. Trades are not executed until both parties deliver. People develop methods of contracting that commit both parties to specifying terms of their agreement and then doing what they agreed to do. Parties aim to do repeat business, developing long-term relationships based on mutual trust. They aim to create an environment where defection is not a dominant strategy but where, on the

[2] The voluntary contributions mechanism is one among many forms that a PD can take. What is necessary and sufficient for a matrix of payoffs to be that of a PD is that the payoffs of the four possible outcomes are in this descending order: unilateral defection > mutual cooperation > mutual defection > unilateral cooperation. A 2x2 PD payoff matrix can be understood as a conjunction of free rider and assurance problems, and psychologically as a conjunction of problems of greed and fear (Isaac, Schmidtz, and Walker 1989). These motivations can be separated into observationally distinguishable components in the laboratory (Schmidtz 1995).

contrary, defecting would poison one's own well and would risk leaving a defector without customers, without suppliers, and so on.

The next section surveys the results of prior laboratory studies of Prisoner's Dilemma-like games, with an emphasis on recent experiments that permit endogenous formation of groups and links. The third and fourth sections describe a new experimental design that will allow us to study the impact of voluntary association where partners decide not only whether to play with each other but also how much to invest in their partnership.

13.2 Experimental work on endogenous group formation

Since the 1980s, a literature has emerged on the effect of being able to choose one's partners. The story begins with early experiments in sociology and political science that considered what happens when subjects have the option of refusing to play. John Orbell and Robyn Dawes (1993) reasoned that introducing an exit option would reduce cooperation because cooperators, victimized by defectors in a mixed population, presumably would be more likely to opt out. However, the opposite effect was observed. Cooperators were *less* likely to exit. The authors concluded that the same factors that tended to make some people more cooperative would make them more willing to continue playing the game.

What happens in *n*-person groups where membership is porous, allowing free entry and exit? Even negative results have been instructive. The problem with free entry and exit is that there is no way to exclude free riders. In the oft-cited Ehrhart and Keser (1999) working paper, subjects were placed in a setting that was analogous to an open access commons.[3] Free riders were free to "chase" high contributors. An influx of free riders would dilute cooperators' productivity. The resulting frustration might induce more cooperative participants to exercise their right to exit.

Are subjects more cooperative when they become a group by mutual consent? For two-person Prisoner's Dilemmas, Hauk and Nagel (2001) compare a *mutual consent* rule (requiring that both players consent to play) to a *unilateral choice* (where either player can force the other to play) rule.

[3] Elinor Ostrom documents the phenomenon of stable communal ownership. See also Ellickson (1993) and Schmidtz and Willott (2003). The most key feature of successful communes is that they find ways to exercise a right to exclude and to distinguish between insiders and outsiders. Successful communes are not open access commons.

Surprisingly, unilateral choice yielded slightly higher cooperation rates (58 percent) than mutual consent (51 percent).

The general result remains, however: an exit option, however structured, results in higher cooperation rates relative to a no-exit baseline.[4] In most economic interactions, exit does not occur in a vacuum; it is accompanied by a switch to another trading partner. Davis and Holt (1994) report results of an experiment with three-person groups composed of one buyer who purchases from one of two potential sellers in each round. The selected seller can either deliver a high or low quality product. Are sellers more likely to deliver high quality when the market is competitive rather than monopolistic? Indeed they are. Cooperative (high quality) outcomes were more than twice as likely in repeated (ten-period) interactions in which buyers could switch to another seller after a low quality delivery. Switching was observed even in two-period variations of this game, but the short horizon was not enough to reward reputations, and in consequence the rate of high quality delivery was no higher than what was observed in one-period controls.

Is there more to being able to choose a partner than simply having a right to walk away? The impact of partner choice is illustrated by Riedl and Ule (2013). Subjects played a series of sixty Prisoner's Dilemmas with randomly changing partners, and with information only about a partner's most recent decision to cooperate or defect. In one treatment, subjects were forced to play. In another treatment, they had an exit option. In a third treatment, subjects were grouped in triples and could list each of the other two partners as acceptable or not. Cooperation in this treatment was over 50 percent, as compared to the low average of about 10 percent in the other treatments. Having a choice among prospective partners fosters cooperation in long sequences of Prisoner's Dilemma games, as compared with a simple exit option.

What happens when players have opportunities to develop reputations that enable them to compete for partners in future games? A stream of papers in biology and ecology studies the impact on cooperation rates when a Prisoner's Dilemma is followed by an opportunity to play a second game where high cooperation can be selectively rewarded. For example, Barclay (2004) used a public goods game in the first stage, followed by a trust game where participants can make leveraged transfers to another subject, who may reciprocate. Similarly, Sylwester and Roberts (2010) used a generalized two-person

[4] A right to say no and walk away from deals one finds unsatisfactory is the hallmark of consumer sovereignty in market society (along with the attendant right of sellers to walk away from offers they deem inadequate).

Prisoner's Dilemma in the second stage with partner selection. The sharpest increases in first-stage cooperation, however, were observed by Barclay and Miller (2007) in a three-person setting where two people interact in the first stage, while the third person observes before choosing which of the first two subjects to interact with in the final stage. Cooperation is much higher in the initial stage, when subjects know that their history will be observed by the third person who chooses a partner in the second stage. Here again we see the influence of a threat of *exclusion*.

What if the competition for future partners is mediated not by endogenous choice but by mechanical sorting? One important discovery is that cooperation rates are higher and decay more slowly when participants are subject to being sorted by the experimenter into more and less cooperative groups. Gunnthorsdottir, Houser, and McCabe (2007) report a voluntary contributions experiment with groupings determined by the experimenter on the basis of current round contributions. The four highest contributors were placed in one group, the next four highest in a second group, and the remaining subjects in the third group. In effect, there was a "meta-payoff" of cooperation, measurable in terms of the quality of the group to which one would belong in future rounds.

Subjects were not told of this procedure but only that assignments to groups would be made after contribution decisions had been made. The baseline control treatment involved random re-sorting after each round. One question crying out for further study: Would contribution levels be affected by *informing* subjects of the nature of the sorting process, thereby giving subjects a measure of conscious endogenous control over the quality of their future partners?

Gunnthorsdottir et al.'s main result is that exogenous contribution-based sorting yielded average contributions higher than random rematching and more resistant to decay. Sorting by contribution level insulated conditional cooperators from the corrosive effects of being matched with significant numbers of free riders. Gunnthorsdottir et al. conclude that almost all of the observed decay in contributions over time was due to reductions by those who were categorized as cooperators, and that rates of decay among cooperators were a function of how frequently they interacted with (and played Tit-for-Tat with) free riders.

The Gunnthorsdottir et al. sorting mechanism neatly separates the two ingredients of successful reputation-building: how one has conducted oneself and how successfully one conveys information to that effect. In Gunnthorsdottir et al., the second consideration was irrelevant, since the sorting process was mechanical. Although this experiment involved exogenous group selection by the experimenter, the result suggests that any process that tends to sort by contribution levels should enhance overall contribution rates. For example, Croson, Fatás, and Neugebauer (2006) use an exogenous rule to exclude the person who contributes the least from the group payoff in a four-person public goods game, which tended to increase contributions.

Another set of papers uses endogenous ranking procedures to sort subjects for subsequent public goods games (Page, Putterman, and Unel 2005, Karmei and Putterman 2013, and Bayer 2011). In the treatment conditions, participants could view others' past contribution histories and submit ordinal preference rankings over possible group partners. The experiments indicate that high contributors are highly ranked as preferred partners. Therefore, these procedures tend to sort on the basis of observed contribution levels, producing a protective sorting effect that enhances overall contributions rates.

Ahn, Isaac, and Salmon (2008) found a mixed effect on contributions levels of allowing subjects to vote on whether to allow entry or exit. Voting to block high contributors from leaving could discourage high contributors. Needing permission to enter could, on the one hand, inhibit the formation of large efficient-sized groups in public goods games, but does, on the other hand, tend to protect high contributors. The strongest effects of voting occur when the group can vote to expel members. In Cinyabuguma, Page, and Putterman (2005), after group members saw histories, a majority vote could expel a member. Vote totals were made public, so the number of "expel" votes served as an effective warning. The expel option was used sparingly, typically with 1–4 people getting expelled from each group of 16. On average, remaining group members contributed over 90 percent of their endowments, with high and steady contribution rates that declined sharply to the baseline level of about 10 percent in the final round. Maier-Rigaud, Martinsson, and Staffiero (2010) also consider the effects of expulsion in iterated public goods games. Expulsion by majority vote was permanent, and resulted in relatively high contributions in the middle periods (about 55–70 percent of endowment even including the zero contributions of expelled members) as compared with 30–50 percent in an analogous public goods game without expulsion. Charness and Yang (2014) provide evidence that expulsion votes are effective when groups start small and expulsion can be temporary, which permits redemption as low cooperators learn to alter their behavior. Expulsion is also effective as a form of punishment when administered unilaterally by a "leader," who makes the initial contribution in a public goods game (Güth et al. 2007).

One clear result of experiments is that cooperation rates are enhanced when links form endogenously. Fehl, van der Post, and Semmann (2011) employed a design in which each subject played independent two-person Prisoner's Dilemmas with several partners, iterated for thirty rounds (with an endpoint not announced in advance so as to avoid endgame effects). In the *static network*, each person was linked with three of ten participants, and links remained fixed for all rounds. In the *dynamic network*, participants were allowed to update one partner link per round, by indicating a desire to continue or not. If one decided to break a link, both were randomly matched with other unmatched participants. Even in the first period, the cooperation rate was higher in the dynamic network (60 percent versus 48 percent), and this difference was approximately maintained during the thirty-round sequence. There were no clusters in the sense that no linked players shared a partner, but clusters of cooperators were common as the dynamic network evolved. Individuals tended to break links with defectors, regardless of whether or not they themselves were reliable cooperators. (See also Rand, Arbesman, and Christakis 2011, Wang, Suri, and Watts 2012.) Note that the breaking of a link in these experiments is unilateral, but the forming of a new link requires mutual consent. In general, cooperation rates are higher in more fluid networks: more frequent opportunities to provide feedback, and opportunities to provide feedback to more partners per period facilitates the processes of cooperating, rewarding, and excluding.

The overall picture that emerges from these literatures is that endogenous group formation in controlled experiments can have large impacts on observed levels of cooperation, even in Prisoner's Dilemma settings where single-period payoffs viewed in isolation seem to dictate defection. Competition for selection into groups is particularly effective when direct exclusion is possible. Such exclusion prevents "cooperation chasing" by free riders. Competition for selection into groups is more effective when interactions are repeated in an informational setting that permits reputation building to be rewarded by subsequent inclusion in profitable groupings. Finally, any process that tends to sort groups by cooperation levels will enhance cooperation when conditional cooperators can exclude free riders. The flip side of this result is that an absence of protective sorting can dampen cooperation.

13.3 Experiment procedures

One consistent theme of existing literature is that being able to exclude non-cooperators has a critically important impact on rates of cooperation.

For individual subjects, being able to exit and to search for more cooperative partners facilitates cooperation. The effects of being able to grow endogenous connections are pronounced when partners can make long-run cooperation more rewarding than long-run defection by making cooperation a condition of lasting access to high quality partners.

We seek to further the experimental work on this topic. This section describes several hypotheses and designs for testing these hypotheses. Our guiding general hypothesis is that standard experimental models and theoretical constructs of the Prisoner's Dilemma systematically underestimate human ingenuity when it comes to ways of circumventing the free rider and assurance problems that jointly define a Prisoner's Dilemma. People have ways of learning to see free riders coming, and of limiting the damage that free riders do. In general, partners do not simply happen to us; we choose them. First, people gather information about who can be trusted and who cannot. Second, people use that information in choosing partners in the first place. Third, people learn from their own experience with particular partners, and decide accordingly whether to maintain or exit the partnership.

Our work is at a preliminary stage, and our results are intriguing but no more than suggestive at this point. First, we surmise that subjects would rather proceed cautiously with new partners. We wondered what would happen if subjects had the option of starting out slowly with a new partner by initially playing for relatively small stakes, then ramping up the scale of risk and reward as trust is established. Would there be any general impact on how successful people are at building enduringly cooperative partnerships? To test this, we introduced a variable (and endogenously chosen) *scale*.

Second, we surmised that subjects would, given the option, prefer to walk away from partners who defected on them, and instead seek new partners. To test this, we incorporated an option of exiting a partnership, then observed how frequently subjects would break off a partnership with a partner whose previous move was defection, versus how frequently subjects broke off relationships with a partner whose previous move was cooperation. Upon breaking off a relationship, how willing would a subject be to risk a new partnership with a randomly selected partner? How cooperative would that new relationship be?

Third, how would the random selection of new partners compare to a setting in which a subject could shop for partners from a pool of subjects whose past record of cooperation was available for inspection? We have not yet tested this hypothesis. (Another extension would be to consider asymmetric settings in which subjects are given two roles and those on one "side" of the transaction, e.g. sellers, have histories that are more visible.) Moreover, in our design, the random reassignment of partners is not altogether random. As subjects could guess, our pools of available partners tended increasingly with the passing of periods to consist of subjects who were available because their previous partnerships had come undone. What subjects could not tell was whether particular partners were available due to their own defection or to their breaking off a partnership with a defector.

Fourth, if scale were an endogenous variable, and if knowledge of a prospective partner's history, or lack thereof, were a variable, how would scalability of trust affect the rapidity with which new pairs of partners would be able to ramp up their partnership to a level of maximal cooperation at maximal scale? This too is work still in the planning stage.

The experiment we describe here allows for endogenous determination not only of access to trading partners but also of the *scale* of trade in a dynamic model of Prisoner's Dilemmas with opportunities for exit and new link formation. One clear result is that setups where links last longer tend to yield higher rates of cooperation. This duration-based difference in cooperation is apparent even in the first round of decisions after a new connection is formed, which suggests that subjects anticipate the benefits of building trust. Unilateral exit is an option in all treatments considered, but when subjects have an opportunity to form new connections in the chosen-partners setting, they tend to start with a low scale and ramp up as trust evolves.

We retain the practice of having a fixed, known final period, so that selfish, rational responses to the incentive to defect in the final period would induce exit in the final round, and then (at least in theory, by backward induction) in all rounds. We limit each person's information to direct observation of decisions by their own partners. (The effect of a broader information base will, however, be something to follow up in a future experiment.) Experimental evidence surveyed in the previous section indicates that generalized reputations about third-party interactions tend to promote cooperation.

The basic game used in our experiments is shown in Figure 13.1, except that we used neutral labels, A and B, rather than Cooperate and Defect. We normalized mutual cooperation's payoff at $1. Payoffs for mutual defection were specified to be negative (−0.50) so that players would prefer to exit from the game and earn $0 insofar as they expected mutual defection otherwise. The row player's payoff is listed first in each cell of the payoff table. For example, if the column player is expected to cooperate, the row player could expect a payoff increase from $1 to $2 by defecting, as indicated

	Column player:	
Row player:	Cooperate	Defect
Cooperate	1, 1	−1.5, 2
Defect	2, −1.5	−0.5, −0.5

Figure 13.1 A Prisoner's Dilemma (row's payoff, column's payoff)

by the bottom left cell of the table. In fact, the payoffs have the property that the marginal incentive to defect is \$1, regardless of whether the other player is expected to cooperate or defect. Such short-term gains, however, could come with diminished opportunities for profitable interactions in the future in a setting where victims of defection could walk away. This potential benefit from competitive altruism is a primary focus of the experiment to be reported.

Even though the experiment was conducted without extra context or "backstory," there is a simple economic exchange interpretation that would generate the payoffs shown in Figure 13.1. For example, suppose the game involves a transaction in which each player delivers a product that the other person values. At a cost of 2, each can deliver a product that the other values at 3, which results in the (1, 1) cooperative payoffs. With a lower cost of 1, however, either player can defect and deliver a low quality product worth only 0.5 to the other, which generates the −0.5 payoffs from mutual defection. A person who defects and incurs the low cost of 1 but receives the high quality delivery from the other person would earn 3 − 1 = 2, where the person who made the high quality delivery and only received low quality in exchange would only earn 0.5 − 2 = −1.5, as indicated in the lower-left box of the table. Finally, the exit payoffs of 0 result from a refusal to trade.

Each session involved 9 rounds and 12 participants, with new binary connection(s) being offered in each round. The treatments altered the numbers of rounds that connections could remain active and the numbers of simultaneous connections that could be activated (details to follow). At the beginning of each round, each participant would view a list of one or more connections, showing the ID number of the other partner and the past history (if any) of prior decisions *with that participant.* Decisions involving other participants were not visible. In many economic exchange settings, the parties to an exchange can control exchange quantities. So, the round began with subjects independently proposing a scale (1x, 2x, 3x, or exit) for each connection. Choosing to exit would in effect be choosing 0x. The actual scale adopted would be the minimum of the respective proposals and would increase from there only by mutual consent. The payoff numbers in the tables would be adjusted to reflect adopted scale for each connection, at which point the actual decisions, A or B, were made for each active connection. After decisions were entered and confirmed, participants would receive earnings results for each connection and they could view a table summarizing all recorded decisions and payoffs for the current and previous rounds for each of their connections, with inactive connections grayed out. An inactive connection in one round could be reactivated if both paired participants agreed to a positive scale in the next round.

If a connection lasts for only one round, then the interaction can be analyzed as a single-round game. For any single-round subgame with a positive payoff scale, the Nash equilibrium is for both to defect. Therefore, payoffs are maximized at 0 by choosing to exit in a single-round interaction. Given the diverse motivations and behavior patterns of human subjects, there will always be some cooperation even in this severe setting. To assess this effect we ran a treatment in which each connection lasted only for a single round, with each person assigned to a new partner in each successive round. This "strangers" treatment is analogous to the one reported by Andreoni and Miller (1993), except that players have the exit option in addition to the usual cooperate and defect options.

The dynamic structures for the various treatments are summarized in Table 13.1. The top row shows the setup for the "strangers" sessions, with

Table 13.1 Sessions and link structure

Structure (number of sessions)	Link duration	Introduction of new partners per round	Active links per round
Strangers, 1 round duration (3 sessions)	1 round with exit	1	1
Chosen partners, short duration (1 session)	$\leq$ 4 rounds with exit	1	3 by choice
Assigned partners, full duration (3 sessions)	$\leq$ 9 rounds with exit	1 in each of 1st 3 rounds	3 by design
Chosen partners, full duration (3 sessions)	$\leq$ 9 rounds with exit	1	3 by choice

connections lasting only a single round. The single-round incentive for unilateral defection in any active connection is $1, $2, or $3, depending on the payoff scale. Based on prior work, we expected low rates of cooperation, especially in later rounds in a session with one-shot interactions. Therefore, we included sessions with dynamic, continuing interactions and opportunities for forming new connections. The basic setup allows subjects to start with a single connection and expand the number of connections over time, a slow growth approach that has produced higher levels of cooperation and coordination in prior experiments of Weber (2006) with coordination games and Charness and Yang (2014) with public goods games. In all of the sessions with enduring connections, each person started with a single connection in round 1, adding a second connection to the first one in round 2, adding a third connection in round 3, etc. In one of the sessions, shown in the second row of the table, the connection duration was limited to four rounds, after which the connection would be automatically deactivated. Thus each person's first round connection would persist for rounds 1, 2, 3, and 4, and each person's new second round matching would be available for rounds 2, 3, 4, and 5, and so on. The Davis and Holt (1994) results for three-person choice-of-partners games strongly suggests that cooperation would be higher in games of longer duration.[5]

The third row of the table describes a setup with connections established sequentially in the first three rounds last through the final round (9), but no new matchings are presented after the third round. In this case, a participant who encounters defection can exit, but it is not possible to encounter another partner to take the place of a deactivated partner.

The final row of the table shows the setup for sessions in which partner selection was possible, since one new matching was presented for each player in each round until the end. To be comparable with other sessions, each person was limited to three active connections in any given round.

The experimental sessions were fairly short: nine periods in total. We did not use practice periods with real or imagined partners because such practice may influence decision-making in early periods. In lieu of practice, participants completed a single-page instructional quiz before starting the

[5] Recall that Davis and Holt did not observe enhanced cooperation in two-round games, but they observed much higher cooperation in ten-round games. The games used in that paper, however, had the property that cooperation could be enforced by a stay or switch punishment strategy. In contrast, the subgame-perfect equilibrium for our multi-round treatments involves no cooperation.

experimental rounds. The quiz was designed to detect misconceptions about the setup: the interactive nature of payoffs, scale choice, the exit option, the duration of connections, and the mutual consent of continued connections. Each quiz was checked, and misconceptions were addressed with each participant before the interactions began.

Participants received a fixed payoff of \$5–12 plus half of what they earned in the experiment, except in the Strangers session, where the fixed payoff was raised to \$16 to adjust for the fact that each person had at most one active connection in each round. Overall, participants earned between \$16 and \$49 for a session lasting about an hour.

This chapter reports preliminary data from 10 sessions, with a total of 120 participants who were recruited from the University of Virginia's subject pool. The experiment is part of a suite of experiments available for use by researchers and teachers on the Veconlab website (http://veconlab.econ.virginia.edu/). The particular program is Networking Games and can be found under the Games menu.

13.4 Cooperation and observed partner selection in the lab

Our first observation is consistent with the widely known tendency for "strangers" in a sequence of one-shot games to defect more often than "partners" who interact in an ongoing series of Prisoner's Dilemmas. Figure 13.2 shows cooperation rates on active links, averaged over sessions. The dark lines at the top pertain to links that endure until the final round, either with chosen partners (from those that are available from the new link presented in each round) or with assigned partners (from those that are presented sequentially in the first three rounds only). With connections that have full duration, average cooperation rates are slightly higher with chosen partners than with assigned partners (although one of the sessions with assigned partners showed more cooperation than one of the sessions with chosen partners). Cooperation rates for both of these partner treatments with longer link duration were higher than for the sessions with links that lasted only 4 rounds or 1 round, as shown by the dashed lines in the bottom part of the figure. The dashed line for the sequence of single-round games shows the lowest levels of cooperation for all rounds. These results for our games with exit opportunities and payoff scale options are consistent with earlier observations that cooperation in a simple Prisoner's Dilemma is higher with assigned partners than with a sequence of new partners (Andreoni and Miller '1993).

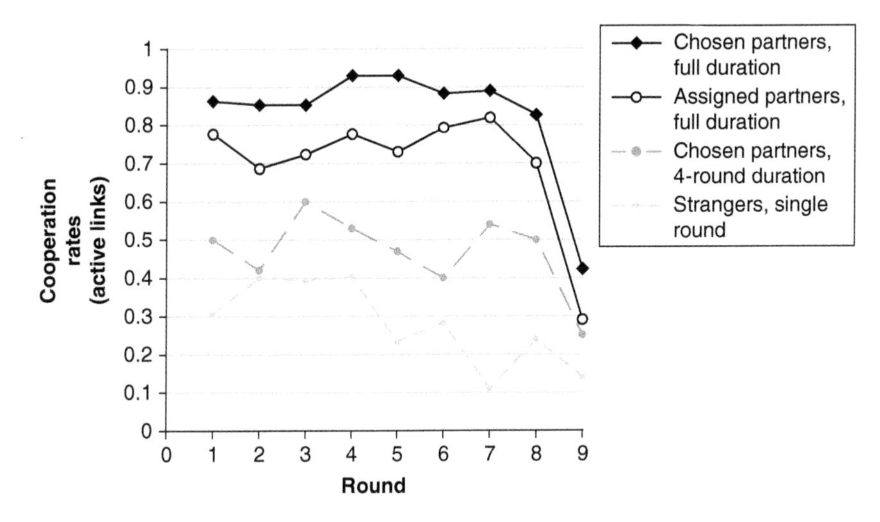

Figure 13.2 Cooperation rates: Impact of choice and of link duration

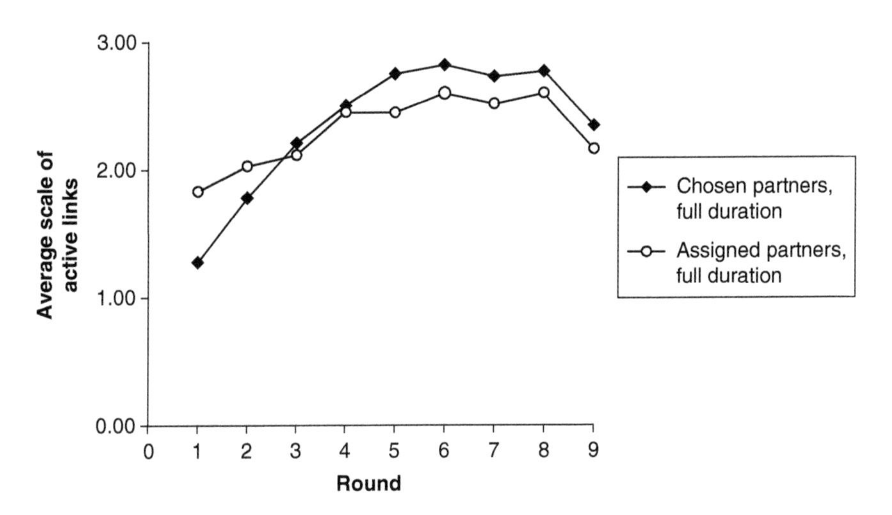

Figure 13.3 Average payoff scales for active links with chosen and assigned partners with full duration

Figure 13.2 suggests that the pro-cooperative effects of longer link duration are more important than the effects of being able to form new links with chosen partners. One major difference between the chosen and assigned partners treatments with full duration is that there is more opportunity to "test the waters" with a low scale when one new prospective partner is encountered in every round. Figure 13.3 shows that, on average, the payoff scales did start lower and rise more in subsequent rounds of the chosen partners setup, as compared with the flatter trajectory of average payoff scales observed with assigned partners.

There is little incentive to try to revive an inactive link when new prospective partners appear each round. In fact, reactivation was never observed in the chosen-partners treatment. The situation is different in the assigned-partners setup, where any inactive link means one less opportunity for a mutually beneficial connection. Therefore, it is not surprising that some people did revive inactive links in the assigned-partners treatment: 28 percent of connections that were severed. Reactivated links, however, almost always exhibited defection by one or both partners. Such reactivation, for example, is observed in the sequence shown in Table 13.2, involving ID2 and the three other subjects who were matched with this person in an assigned-partners session. The top row is a typical example of a profitable pairing starting at 1x scale, with scale increases following mutual cooperation (both players choosing decision A). Notice that the other player (ID7) defected in round 8, and the resulting exit by one or both players caused the link to be inactive in the final round. The second row shows a matching ending in defection by the other player in the very first round, at 2x scale, with the link remaining inactive after that. This dead link illustrates why the ability to exit is not sufficient to generate sustained high rates of cooperation. The third row of the table shows a link that became inactive after an initial defection and was reactivated in round 5, but the subsequent level of trust was tenuous. This attempt to cooperate was unusual in the sense that the most common outcome after link reactivation involved immediate mutual defection.

One interesting feature of the patterns in Table 13.2 is that a defection in the very first interaction for a given link is a key determinant of subsequent payoffs for that link. For example, the other player defected in the initial

Table 13.2 Interaction history of decisions for ID2 (listed first) with decisions for three assigned partners (listed second) and with the scale (1x, 2x, or 3x)

Round:	1	2	3	4	5	6	7	8	9
ID 2 with ID 7:	A A 1x	A A 2x	A A 3x	A A 3x	A A 3x	A A 3x	A A 3x	A B 3x	Inactive
ID 2 with ID 8:		A B 2x	Inactive	Inactive	Inactive	Inactive	Inactive	Inactive	Inactive
ID 2 with ID 12:			A B 1x	Inactive	B A 1x	A B 1x	A A 1x	A B 1x	B B 1x

round for two of the three rows in the table. One possibility is that initial defections might be *more* common in the chosen-partners treatment than with assigned partners, since a soured relationship is not so costly if new partners appear in each subsequent round. A counter-argument would be that initial defections should be less common in the chosen-partners treatment since it may be harder to find reliable partners in later rounds, insofar as cooperators might tend already to have their maximum of three reliable partners and thus not be in the market for new partners. Neither of these conjectures is confirmed by the small sample of data currently in hand: The percentages of defect decisions in the initial round of an active link (18 percent with chosen partners and 31 percent with assigned partners) are not statistically different using a Wilcoxon two-sample test with individual initial defection rates as observations. Initial defection rates were significantly higher (again based on a Wilcoxon test) for subjects facing shorter links (50 percent for four-round duration and 61 percent for single-round duration), although there is no significant difference between the four-round and single-round initial cooperation rates.

Successful interaction involved both cooperation *and* having a high proportion of possible links activated at high scales. One measure that takes cooperation, scale, and link activation into account is overall efficiency, which is measured as actual earnings (net of fixed payments) as a percentage of the earnings achievable with the maximum permitted number of links operating at 3x scale with mutual cooperation. If a person has one link in the first period, two in the second, and three links in each of the next eight periods, there are twenty-four possible links, with a resulting maximum earnings of 3x24 = $72.[6] In contrast, there are only nine possible links with a sequence of single round matchings, so the maximum earnings used in the efficiency formula would only be 3x9 = $27 in that case. Table 13.3 shows summary performance measures, including efficiency, averaged over all sessions in each treatment.

Cooperation rates on active links are shown in the left column of Table 13.3. They increase as link duration goes up. The next column shows percentage of the possible links that were active. These were similar for three of the rows and somewhat lower with assigned partners. This difference is due to the fact that exits following defections could not be replaced with new links after the third round. Average scales for active links, shown in the third column,

[6] Recall that actual earnings were divided by 2, but this does not matter for the efficiency calculations.

Table 13.3 Performance measures and link structure

Link structure	Cooperation rate (on active links)	Active links (pct. of possible links)	Avg. scale (on active links)	Efficiency (earnings as pct. of maximum)
Strangers (1-round duration)	29%	75%	1.9	–3%
Chosen partners (4-round duration)	47%	83%	2.2	11%
Assigned partners (full duration)	71%	69%	2.3	35%
Chosen partners (full duration)	84%	80%	2.4	52%

are also similar for all rows, although the summary numbers in the table do not reveal the pattern across rounds shown earlier in Figure 13.3. In particular, the "hump-shaped" slow buildup of scale with chosen partners is not observed in the other treatments. Efficiencies in the final column diverge sharply. Total earnings for the sequence of one-shot Prisoner's Dilemma games were minimal, as losses due to defections tended to negate gains, with a resulting efficiency near zero (and negative in two sessions). Efficiencies are substantially higher when links have longer durations, and higher with chosen partners.

Consider the relation between individual earnings and individual strategies. Given that defecting earns a dollar more than cooperating no matter what one's partner does, one would expect to see a negative correlation between cooperating and earnings in a single-round play. This correlation would become positive if the per-round cost of cooperating is coupled with a long-run benefit of maintaining mutually profitable, active links. To see whether this actually happened, we calculated the correlations between individual earnings for the whole session and separate cooperation rates for rounds 1–3, rounds 4–6, and rounds 7–9. These correlations are shown in Table 13.4.

As expected, correlations between earnings and cooperation rates are highly negative for the session with 1-round link duration, shown in the top row of Table 13.4. Correlations are smaller and are not significantly different

Table 13.4 Correlations between individual earnings for all rounds and cooperation rates on active links in each third

Link structure	First part (rounds 1–3)	Second part (rounds 4–6)	Third part (rounds 7–9)
Strangers (1-round duration)	-0.56^{***}	-0.57^{***}	-0.52^{**}
Chosen partners (4-round duration)	-0.25	-0.39	-0.13
Assigned partners (full duration)	0.40^{***}	0.53^{***}	0.35^{**}
Chosen partners (full duration)	0.61^{***}	0.47^{**}	0.47^{***}

Key:
** indicates a significant difference from 0 at $p = 0.05$ (1-tailed test).
*** indicates a significant difference from 0 at $p = 0.01$ (1-tailed test).

from 0 for subjects who had links with 4-round durations. Correlations between cooperation rates and earnings are highly *positive* for the sessions with links that endured until the final round, as shown in the two bottom rows of the table. A review of all subjects' interaction records, analogous to the record in Table 13.2, reveals that the key to high earnings in either of the full duration partners treatments is to cooperate in the initial round and raise scale to the maximum of 3x, holding it there until the final or penultimate round.

In the chosen-partners sessions with full link duration, the cooperative decision was selected *in the final round* more than half of the time, i.e. in 35 out of 64 cases in which links were active in the final round. The person who made the least money in chosen-partners sessions was someone who defected in the initial round, which in every case triggered exit. This person did not achieve the permitted maximum of three active links in most rounds.

13.5 Earning trust, earning esteem

Adam Smith would not be surprised by what we see in the laboratory. To Smith, self-interest was not as fundamental a psychological foundation as some readings of *Wealth of Nations* assume. Strikingly, when Smith opens *Wealth of Nations*, book I, chapter 2, by asking what accounts for the evolution of specialization, his opening remark refers not to the profit motive but to the propensity to truck and barter.[7] This propensity, Smith says, is a necessary attribute of social beings whose ability to cooperate is mediated by reason and speech. It might be grounded in the profit motive or might itself be a primordial human motive.[8] A drive to truck and barter is a drive not only to make money but more fundamentally a drive to make deals. It is a drive to reciprocate favors, cultivate allies, and be part of a community of people who each bring something good to the table – free and responsible reciprocators who warrant esteem and whose esteem is thus worth something in return.[9] A merchant learns how to bring something to a community that makes it a better place to live and work for everyone with whom that merchant deals. The overall result may be no part of a trader's intention, as Smith says in places, but neither is a successful trader's intention simply a matter of self-absorbed acquisitiveness.

It makes perfect sense for the author who treated benevolence as primary in his first book subsequently to analyze market virtue as a matter of treating the self-love of trading partners as primary. As a benevolent person hoping to truck and barter with the brewer and baker, you think first of their self-love because you want them to be better off. Smith does not say bakers are motivated solely by self-love. What he says is that we do not *address ourselves* to their benevolence but to their self-love. This is not to make an assumption about what motivates bakers. Rather, it is to reflect on what it takes to be benevolent oneself in dealing with bakers – on what it is like to be considerate.[10] In sum, the author of *Moral Sentiments* gives center stage to virtue and benevolence, but in elaborating the substantive content of these

[7] The first sentences are these: "This division of labour, from which so many advantages are derived, is not originally the effect of any human wisdom, which foresees and intends that general opulence to which it gives occasion. It is the necessary, though very slow and gradual, consequence of a certain propensity in human nature which has in view no such extensive utility; the propensity to truck, barter, and exchange one thing for another. Whether this propensity be one of those original principles in human nature, of which no further account can be given; or whether, as seems more probable, it be the necessary consequence of the faculties of reason and speech, it belongs not to our present subject to enquire."

[8] Schmidtz thanks Geoff Brennan for a helpful discussion of this point.

[9] Smith (1759), VI.i.3. The desire to be proper objects of esteem may be "the strongest of all our desires" and for good reason. Nowadays, our bodily needs are easily met, whereas the esteem of our peers is a hard-fought daily battle that on any given day may hold the key to our fate.

[10] Although this does not posit self-love as primary, it does invite reflection on the fragility of all motivations, self-love and benevolence included. Benevolence needs nurturing. One way to nurture it is to avoid leaning too hard on it, and celebrate when it culminates in flourishing rather than self-sacrifice. To Smith, self-love likewise needs nurturing, and our failure to keep our true interest in focus is lamentably common.

ideas, the author of *Wealth of Nations* notes what should be obvious: namely, a man of true benevolence wants his partners to be better off with him than without him.

The harmony of interests among free persons is not remotely to be taken for granted, yet is manifestly a real possibility. So long as people can see a way of building a community of partners who are better off with them than without them, and so long as they see themselves as having reason to cherish such an achievement, their self-interest will bring them together to form a free and thriving community.

Laboratory experiments provide insights into the effects of endogenous networking in Prisoner's Dilemmas. Agents seek relationships with those who have a record of being cooperative. But it is not only cooperators who seek out cooperators. Free riders do the same. Thus, letting people sort themselves into groups is not enough if free riders can "chase" cooperators.

We encourage people to be magnanimous, but what honestly encourages people to be magnanimous is putting them in a position where they can afford to believe in each other – where they are not at the mercy of "partners" who may be less than magnanimous. When people are free to make their own decisions about whom to trust, and when they have some liberty to exit from relationships that go sour, they will be more free to enter relationships (that is, more able to afford the risk) in the first place. The Prisoner's Dilemma is the structure of many of our fundamental challenges as social beings. But we have ways of coping. We learn when exit can be and should be a unilateral decision. We learn how to exclude (and thus when *entry* can be and should be a *mutual* decision). We evaluate. We gossip. We learn to treat reputation as the coin of the realm.

Even though defection always results in a higher payoff, there is a positive correlation between individuals' cooperation rates and earnings in the session. The "dark side" of this process is that, once trust is gone, there is no easy way to rebuild. In our experiments, links are never reactivated in the chosen-partners treatment after a player decides to exit. With the assigned-partners treatment, long duration links are sometimes reactivated, but post-reactivation cooperation was invariably problematic.

The long duration treatments generate high and sustained rates of cooperative choices for active connections. Such cooperation sometimes continued into the final round, although overall economic efficiency is diminished by frictions associated with inactive links, transitions, and some endgame behavior. The observed high rates of sustained cooperation with full duration links are notable, given that: 1) "long" is only nine rounds, not infinite horizons with discounting; and 2) individuals cannot observe the histories of others' decisions involving third parties. Hence, each new link opportunity in the experiment involves trusting a "stranger." One interesting extension would be to evaluate the effects of the public posting of subjective ex post-performance reports made by parties to a transaction. Thus, a laboratory design that makes a subject's history transparent, and enables subjects to select partners based on such histories or the basis of peer-review ratings, could be a fairly realistic representation of the vehicles for reputation-building that we see in society at large. Thus, there is potential in these designs for testing policy as well as for testing theory.

14 The lesson of the Prisoner's Dilemma

Paul Weirich

The Prisoner's Dilemma teaches many lessons about individuals interacting. A very prominent lesson, the one I treat and call its lesson, concerns standards of rationality. This lesson reveals profound points about the relationship between rationality's standards for individuals and its standards for groups.

14.1 Rationality

Rationality is a normative and evaluative concept. Agents should act rationally, and their acts, if free, are evaluable for rationality. Rationality considers an agent's abilities and circumstances before judging an act. Because rationality recognizes excuses, acting irrationally is blameworthy. If an agent has an excuse for a defective act so that the act is not blameworthy, then it is not irrational. These points pertain to the ordinary, common conception of rationality, which I use, rather than a technical conception that offers precision but has less normative interest.

A general theory of rationality with explanatory power covers rational action in possible worlds besides the actual world. A *model* for a theory of rationality constructs a possible world with features that control for factors in the explanation of a rational choice. For instance, a model may specify that agents are fully informed despite agents' ignorance in the actual world. A model's assumptions may be either idealizations or just restrictions. An idealization states a condition that promotes realization of goals of rationality, which rational agents want to attain, such as making informed choices. Because an idealization facilitates attaining a rational agent's goals, rational agents want its realization too. For example, because a rational agent wants to make informed choices, she also wants to gather information. Full information about relevant facts is thus an ideal condition for a decision problem. In contrast, a mere restriction states a condition that does not promote goals of rationality but may simplify explanations of rational action. For example, that an agent has a linear utility function for amounts of money is a restriction rather than an idealization because a rational agent need not want to have such a utility function either as a goal or as a means of attaining a goal of rationality.

Knowledge of the choices that are rational in a model guides rational choices in real decision problems that approximately meet the model's idealizations and restrictions. In addition, principles that explain rational choices in a model may offer partial explanations of rational choices in the actual world by describing the operation of some factors in full explanations.

A theory of rationality begins with standards of rationality for individuals and then considers their extension to groups. A common proposal for individuals in a decision problem is the requirement to maximize utility, that is, to adopt an option with utility at least as great as the utility of any other option. Some theorists, such as Allingham (2002: chapter 2), define utility using preferences so that by definition choices maximize utility. However, so that maximization of utility is a norm, I take utility to be rational strength of desire and take an additional norm to require that preferences agree with utilities. Given satisfaction of the second norm, maximizing utility agrees with selecting an option from the top of a preference ranking of options.

Maximization of utility is a goal of rationality. A rational individual wants to maximize utility because doing this realizes other desires as fully as possible. However, impediments to maximization, such as time pressure, often excuse failures to maximize. Consequently, maximization is not a requirement in all cases but is a requirement if conditions are ideal for maximization. In general, rationality requires only promotion of utility maximization, either by achieving it or by taking reasonable steps to reach a position in which one may approximate it, for example by reviewing and evaluating options when this is easy.

Although necessary for rational action in ideal conditions, maximizing utility is not sufficient for rational action if an agent has irrational goals and so an irrational assignment of utilities to acts. Then maximizing utility need not generate a rational act. It may generate an act that is instrumentally but not fully rational.

Groups act, and their acts are evaluable for rationality when free. Collective rationality governs collective agents. Because the ordinary concept of rationality extends to collective agents, collective rationality does not require a technical definition. I do not define collective rationality, for example, as (weak) Pareto-efficiency, that is, performing a collective act such that not all group members alike prefer some alternative. By not defining collective rationality as efficiency, I let the collective rationality of achieving efficiency remain a normative issue.

How do standards of rationality for individuals extend, if they do, to groups? The straightforward extension to groups of the standard of utility maximization requires the maximization of collective utility, suitably defined. The traditional definition of an act's collective utility makes it a sum of the act's interpersonal utilities for the group's members, that is, the sum of the act's personal utilities put on the same scale. Some definitions propose substituting the act's average interpersonal utility to accommodate groups of variable size. However, to put aside this issue, I treat only groups with constant membership. For such groups, defining collective utility as total interpersonal utility works well.

Because information influences personal utility, it also influences collective utility. As Broome (1987) explains, when the members of a group do not have the same information, collective utility defined as total utility ranks collective acts as does Pareto-superiority, but may not be Bayesian (so that an act's collective utility is a probability-weighted average of the collective utilities of its possible outcomes for some probability function over its possible outcomes). To put aside issues concerning collective rationality that arise when a group's members have different information about the outcomes of their collective acts, I treat models in which a group's members are fully informed about the outcomes of the options in their collective action problems.

The moral theory, utilitarianism, advances maximizing collective utility as a requirement of right action, but I investigate whether, in models satisfying suitable conditions, collective rationality requires maximizing collective utility. Individuals often care about morality so that it affects their personal utility assignments and hence a collective act's collective utility. However, I consider arguments for maximizing collective utility that do not rely on the members' wanting to be moral.

In a typical collective action problem calling for division of a resource, every division is efficient, but only one division maximizes collective utility. Commonly, many collective acts are efficient, but few maximize collective utility. Maximizing collective utility is more demanding than is efficiency.

Using personal utility to represent the preferences of each member of a group, efficiency requires a collective act for which no alternative is higher in each member's utility ranking. Some theorists advance for collective rationality only a requirement of efficiency because efficiency's definition does not use interpersonal comparisons of utility. However, a theory of rationality may use interpersonal comparisons, despite their impracticality, to strengthen requirements of collective rationality. A model for the theory may assume that interpersonal utilities exist and then investigate whether given this assumption collective rationality requires maximizing collective utility. The difficulty of measuring interpersonal utilities may hamper the model's application to the real world but does not afflict the model itself.

Maximizing collective utility agrees with realizing a collective act at the top of a collective ranking of options in a collective action problem. Moreover, ranking collective acts is simpler than evaluating each act's collective utility and then using collective utilities to rank acts. Nonetheless, I examine standards of collective rationality that use collective utility rather than collective preference because, as Arrow's (1951) famous theorem argues, collective preference has no adequate general definition.

A group's maximizing collective utility, assuming informed and fully rational personal utility assignments, is good for the group in a common technical sense. However, an argument that collective rationality requires maximizing collective utility must show that a group of rational agents maximizes collective utility. The argument must show that a group's maximizing collective utility is consistent with the rationality of the group's members.

A theory of rationality sets consistent requirements for individuals and for groups. If the members of a group act rationally, then a collective act that the members' acts constitute is collectively rational. If it were not, then rationality would prohibit the collective act despite permitting each member's part in the act. So that principles of rationality do not generate inconsistencies, such as forbidding a group to stand while allowing each member to stand, the principles do not forbid a collective act while permitting the individual acts that constitute the collective act. A collective act cannot change without a change in the individual acts that constitute it. If a group's act should have been different, then some member's act should have been different.

The members of a committee that adopts an ill-conceived resolution bear the blame for the resolution. The blame the committee bears transfers to the members because had they voted differently, the resolution would not have passed. Which members are culpable depends on details of the case. If a group is to blame for an act, then some member or members are to blame for their parts in it.

The transmission of blame from the group to its members is not a matter of logic. It is logically possible that a group is to blame for a collective act although no member is to blame for her part in the act. Although some member should have acted differently, perhaps no member is such that she should have acted differently. The transmission of blame is a normative, not a logical, matter.

Suppose that every member of a group performs an act, and each act is permissible in its context. Then the members' acts taken together are permissible. If the combination is not permissible, then although each member's act may be permissible conditional on some combination of acts, given the actual combination of acts not all acts are permissible. Consider a collective act that a group performs by its members' acting one by one in a sequence. Suppose that the nth act is permitted given its predecessors, the $(n - 1)$th act given its predecessors, . . ., the 2nd act given its predecessors, and the 1st act. Then all acts in the sequence are permitted, so the whole sequence is. If the sequence is not permitted, some act in the sequence is not permitted given its predecessors. If a group is to blame for leaving on the lights after it leaves its meeting hall, then the last member to leave is to blame. Because the rationality of each member's act suffices for the rationality of the collective act that the members' acts constitute, individual rationality does not conflict with collective rationality, contrary to some theorists such as Sen (2002: 212).

Given that collective rationality arises from individual rationality, does collective rationality require maximizing collective utility? If so, then if every member of a group acts rationally in producing a collective act, the collective act, being collectively rational, maximizes collective utility. Collective rationality does not require maximizing collective utility if members of a group that act rationally sometimes fail to maximize collective utility.

14.2 The Dilemma

The Prisoner's Dilemma is a noncooperative game without opportunities for binding contracts. Two people each benefit if both are cooperative instead of uncooperative; however, each benefits from being uncooperative no matter what the other does. To settle comparisons of collective utilities, I assume that the total utility if both act cooperatively is greater than the total utility if one acts cooperatively and the other acts uncooperatively. Hence both acting cooperatively maximizes collective utility. Table 14.1 lists in the cell for each combination of acts the combination's interpersonal utility, first, for the row-chooser and, second, for the column-chooser.

In the Dilemma, if each acts rationally, neither acts cooperatively (or so I assume). Because the two fail to cooperate, their collective act is not efficient and moreover does not maximize collective utility. Their failure to cooperate, because it arises from each individual's acting rationally, is nonetheless

Table 14.1 The Prisoner's Dilemma

	Act cooperatively	Act uncooperatively
Act cooperatively	2, 2	0, 3
Act uncooperatively	3, 0	1, 1

collectively rational. The Dilemma shows that collective rationality does not require efficiency or maximizing collective utility in all cases. When conditions are adverse for coordination, as in the Dilemma, collective rationality may excuse failures to achieve efficiency and to maximize collective utility. This is the Dilemma's lesson.

A collective act's irrationality entails its blameworthiness, as for individuals. Extenuating circumstances deflect blame and charges of irrationality for defective collective acts. Inefficiency is a defect, but nonideal conditions for coordination excuse it. In the Dilemma the pair of individuals, if rational, fails to achieve efficiency but escapes blame because conditions impede coordination; the pair cannot enter binding contracts. Suppose that conditions were ideal for coordination, in particular, the two were able to enter a binding contract without cost. Then it would be rational for one to propose a contract binding both to cooperation, and it would be rational for the other to accept the contract. Rationality requires their taking steps to put themselves in position to achieve efficiency, and the contract does this. First, they enter the contract, and then they execute its terms. The result is cooperation and efficiency. In the version of the Dilemma I assume, the result also maximizes collective utility.

It is rational for each individual to enter the contract because she is bound to comply only if the other individual does, and both are bound to comply and will comply if both enter the contract. The contract guarantees that both do their parts, and each gains if both do their parts, so the contract is not a burden for either party. The benefits of executing the contract make entering it rational. Although entering the contract has consequences different from executing the contract, in ideal conditions for coordination, the difference is negligible.

The Dilemma shows that efficiency is a goal of collective rationality rather than a general requirement. A goal of collective rationality is not a goal of a group (assuming that groups lack minds) but rather a goal of each member of the group, if each is rational (and cognitively able to have the goal). Rational members of a group want to attain goals of collective rationality.

Each member benefits from realizing an efficient collective act instead of a collective act Pareto-inferior to it. Their desires prompt organization to achieve efficiency. A member typically cannot achieve efficiency by herself but can take steps to promote efficiency, such as proposing a contract binding all to efficiency.

Corresponding to a goal of collective rationality are ideal conditions for attaining it. Rational members of a group want to realize the ideal conditions because then they are in position to achieve the goal. A rational member takes reasonable steps to put the group in position to attain the goal approximately. Rational agents, if they can, create opportunities for binding contracts so that Prisoner's Dilemmas do not arise. They may impose penalties for failing to act cooperatively that change payoffs and eliminate the Dilemma. "Solutions" to the Dilemma change the game so that acting cooperatively is rational.

Suppose that we define a game using its payoff matrix. Then making conditions ideal for the game changes the situation in which the game is played but not the game itself, provided that the payoff matrix is constant. For the Dilemma, the payoff matrix may be the same given the possibility of costless, binding contracts that take compliance out of the agents' hands, but it is not the same given repetitions of the Dilemma because then payoffs include the effect of a player's current choices on the other player's future choices. Although the possibility of binding contracts is an ideal condition for the game, because repetition changes the game, repetition is not an ideal condition for the game.

A circumspect definition of a game encompasses all features that affect the game's solution, including opportunities for binding contracts. Accordingly, the Dilemma is defined as a noncooperative game, without the possibility of binding contracts, and is not defined just by its payoff matrix. Adding the possibility of binding contracts changes the payoff matrix's context and so eliminates the Dilemma. Thus, the possibility of binding contracts is an ideal condition for coordination but not an ideal condition for the Dilemma itself.

An efficient collective act need not have the support of all agents in a bargaining problem over division of a resource. An unequal division is efficient but has not the support of all if failing to realize it yields an equal division. Some agents gain in the move from an unequal division to an equal division, and even an agent who loses in the move supports the equal division if the alternative is a failure to reach an agreement and, consequently, wasting the resource to be divided. However, for any inefficient outcome, some efficient outcome has the support of all, so collective rationality yields efficiency. Efficiency is a goal of collective rationality and a requirement when conditions are ideal for coordination, as Weirich (2010: section 10.4) argues.

Is collective-utility maximization also a goal of collective rationality and a requirement in ideal conditions for coordination? Does rationality require a group to promote maximization of collective utility either by achieving it or by taking reasonable steps to put the group in position to maximize collective utility approximately, and does collective rationality require a group to maximize collective utility when conditions are ideal for coordination?

To put aside some objections to a group's maximizing collective utility, I make some assumptions. First, I assume that membership in the group is voluntary. If not, rationality may not require a member to participate in maximization of the group's collective utility. An individual may belong to a group of castaways only because she is trapped on an island with them, and she may be opposed to the group and not want the good for it. Second, I assume equal social power. Otherwise, a case may be made that collective rationality requires maximizing power-weighted collective utility, as Weirich (2001: chapter 6) explains. Third, I assume that group members are cognitively ideal and fully rational so that they have rational goals and rational personal utility assignments. Otherwise, the group may have excuses for not maximizing collective utility, and maximizing collective utility may be instrumentally but not fully rational for the group. Fourth, I treat only collective action problems in which a group's collective act arises from the participation of all members, and a member's participation expresses consent. This bars cases in which a group maximizes collective utility by having some members confiscate another member's goods.

These assumptions are not independent. If a member of a group suffers because of collective acts to which she does not consent, she may leave the group if she can. The assumptions also mix idealization and restriction. Rational agents want to have rational goals, want their membership in a group to be voluntary, and want their participation in collective acts to be voluntary. However, it is not clear that rational agents (as opposed to moral agents) want equality of social power in groups to which they belong. So I bill this equality as a restriction rather than as an idealization. The next section shows that a case for maximizing collective utility requires additional restrictions.

14.3 Self-sacrifice

A group's maximizing collective utility, in contrast with its achieving efficiency, need not reward every member. Consequently, maximizing collective utility is not a general goal of collective rationality. In some situations the members of a group, although rational, do not each want to maximize collective utility even if conditions are ideal for coordination and meet the previous section's assumptions.

Consider a case in which a member of a group by sacrificing his own life may complete a collective act that enormously benefits each member of a group so that their gains outweigh his loss. The collective act of which his sacrifice is a component yields a total utility for the group at least as great as the total from any alternative collective act and so maximizes collective utility. Nonetheless, the individual whose sacrifice benefits the group may rationally want to spare himself rather than maximize collective utility. He may rationally resist sacrificing himself for his group. His rational resistance and the rational acts of others constitute a rational collective act that fails to maximize collective utility. Because collective rationality does not require maximizing collective utility in such cases, maximizing collective utility is at most a restricted goal of collective rationality, that is, a goal only in restricted cases. It is not a requirement whenever conditions are ideal for coordination but at most a requirement when, besides ideal conditions, suitable restrictions hold.

Do any interesting restrictions, added to the previous section's assumptions, do the job? Some that suffice include the possibility of compensating members of a group who lose from the group's maximizing collective utility. Then the group, having rational members, uses compensation to remove objections to maximizing collective utility.

Two conditions are necessary for the possibility of suitable compensation. First, an individual who loses in maximization of collective utility must not suffer a loss for which compensation is impossible. His loss cannot be his life, assuming that nothing compensates for this loss. Second, the group must have a means of compensating losers so that the sequence of collective acts consisting of maximizing utility and then compensating losers itself maximizes collective utility. Otherwise, the initial collective act does not lead to maximization of collective utility on balance. Suppose that someone in a group is very good at converting the group's resources into utility; the group maximizes collective utility by giving all its resources to him. Compensating others is possible afterwards; returning to the original distribution of resources compensates them. However, maximizing collective utility and then compensating losers in this fashion does not maximize collective utility on balance.

Suppose that each collective act in a two-act sequence maximizes collective utility. In special cases, the sequence nonetheless may not maximize collective utility. However, in typical cases, if each act in the sequence maximizes collective utility, then the sequence also maximizes collective utility among alternative sequences. To simplify, among two-act sequences with, first, a collective act that maximizes collective utility and, second, a collective act that compensates losers, I treat only sequences that maximize collective utility just in case the compensation step, as well as the initial step, maximizes collective utility. In these cases, the second condition necessary for the possibility of suitable compensation reduces to the availability of a compensation step that itself maximizes collective utility.

Compensation may motivate a group's members to participate in a collective act that maximizes collective utility even if the resulting utility distribution is uneven. It motivates, assuming that compensation considers everything that matters to the group's members, including feeling envy, if this occurs. It gives a member who loses by helping to maximize collective utility an amount of personal utility at least equal to the amount of personal utility he loses by helping. In some cases, because of envy, compensation after maximization of collective utility may be impossible. Restricting maximization of collective utility to cases in which compensation is possible puts aside cases of extreme envy.

Ideal conditions for coordination by themselves do not ensure a mechanism for compensation. The possibility of binding contracts does not remove constraints on the terms of contracts, and contracts do not remove barriers to maximization of collective utility if terms cannot include compensation. Also, a group's having a mechanism for compensation is not itself an ideal condition for coordination. A group's members, even if rational and ideal, need not want, to promote coordination, a mechanism for compensation. Coordination may occur without arrangements for compensation, as it does not require maximizing collective utility. Consequently, a model's stipulating a mechanism for compensation introduces a restriction rather than an idealization.

Utility is *transferable* if it is possible to move utility from one person to another at a constant rate of transfer, perhaps using money or a resource as the vehicle of the transfer. Transferable utility need not be interpersonal; a unit transferred need not have the same interpersonal utility for the donor and the recipient. Transferable utility makes compensation possible. Suppose that utility is transferable among the members of a group, and some collective act maximizes transferable utility. Then it also maximizes collective utility. It is collectively rational to maximize collective utility if the gainers compensate the losers with utility transfers to produce an outcome that is better for all than failing to maximize collective utility.

Compensation that transfers utility from gainers to losers may not preserve maximization of collective utility in some cases. It may transfer utility by moving a resource from someone who uses a little of the resource to obtain a lot of interpersonal utility to someone who uses a lot of the resource to obtain a little interpersonal utility, thereby lowering collective utility. In this case, maximizing collective utility followed by compensation does not maximize collective utility on balance. The previous section's second condition concerning compensation requires the possibility of compensation that does not lower collective utility, for example compensation that costlessly redistributes collective utility. More specifically, the second condition requires the possibility of compensation that maximizes collective utility.

In the model the idealizations and restrictions create, the interactions of a group's members form a cooperative game with maximization and compensation stages. As Binmore (2007b: chapter 9) and Weirich (2010: chapter 8) explain, a *cooperative game* offers opportunities for coordination, and the game's *core* is the set of outcomes in which each coalition of players gains at least as much as it can gain on its own. In a two-act sequence with maximization of collective utility followed by compensation for losers, the compensation ensures that each member, and moreover each coalition of members, does at least as well participating in the sequence as acting alone. The sequence produces a core allocation.

A rational group need not take steps in all situations to facilitate maximizing collective utility because maximization is a restricted goal. However, in the model that collects the restrictions and ideal conditions stated, agents may coordinate effortlessly and arrange compensation that maximizes collective utility. A group may compensate a member who loses by participating in collective utility maximization. In the model, a rational group maximizes collective utility, perhaps using institutions of compensation it has established to facilitate maximization of collective utility. Given any collective act that fails to maximize collective utility, agents have incentives and the means to maximize instead. Collective rationality requires maximizing collective utility.

The argument for the requirement assumes that collective rationality requires efficiency when conditions are ideal for coordination and then notes that the restrictions added make maximizing collective utility necessary for efficiency. If all the efficient collective acts have some property, such as maximizing collective utility, then, collective rationality yields a collective act with that property. With a mechanism for compensation, only collective acts that maximize collective utility are efficient. For any act that fails to maximize, another exists that maximizes and after compensation realizes an outcome

Pareto-superior to the first act's outcome. Maximizing collective utility in a group with a suitable institution of compensation is (weakly) Pareto-efficient because no alternative benefits each member of the group. Moreover, because each member prefers a maximizing act with maximizing compensation to any alternative act that does not maximize, no such alternative is (weakly) Pareto-efficient. Efficiency supports maximizing collective utility given a suitable institution of compensation.

14.4 Mutual aid

How may a group meet the demanding conditions under which rationality requires maximizing collective utility? Consider a diachronic reformulation of the Prisoner's Dilemma in which one agent may act cooperatively first, and the other agent may act cooperatively later, with the possibility of a binding contract to ensure their cooperation. For example, two farmers may contract to help each other with their harvests. The farmer who helps first suffers a loss that the second farmer compensates later. The second farmer's cooperative act is compensation for the first farmer's cooperative act. It brings the first farmer, who initially loses, a gain later. The two should enter a binding contract requiring each to act cooperatively and thereby maximize collective utility.

A diachronic reformulation of the Prisoner's Dilemma, given suitable background assumptions, exhibits *a policy of mutual aid*. Such a policy, as I define it, calls for a member of a group to help another member when the cost to the first member is little, and the benefit to the second member is great. The policy compares gains and losses interpersonally but without precision. In a pair's application of the policy, the first member consents to give aid, and the second member consents to receive aid, so that their participation in the policy forms a collective act. Adherence to the policy maximizes collective utility, assuming that no alternative to the first member's helping the second member creates even more collective utility. In a large group, the consent of all at a time to the policy forms a collective act. Custom or contract may secure the policy. In the first case, a violation brings a penalty so great that no member incurs it; heavy disapprobation and no future help comes to a member who violates the policy. For a group with a policy of mutual aid, given some background assumptions, collective rationality requires maximization of collective utility in cases the policy governs.

Participation in a policy of mutual aid is rational for each member of a group, and so collectively rational, provided that a participant who loses expects to be a beneficiary in the future. The future help expected may come from a member of the group besides the policy's current beneficiary. An expectation, not necessarily a certainty, of future gain may be adequate compensation. It is adequate if the probability of future gain is high enough to motivate a rational agent.

Compensation for participation in a policy of mutual aid is not possible in all cases. A person on his deathbed who helps another now may not expect to live long enough to benefit in the future from others' help. So background assumptions include that the participants in a policy of mutual aid expect to live long enough to receive compensation for providing aid. Given this assumption and others already stated, collective rationality requires participation, which maximizes collective utility. For simplicity, I explain this only for cases in which compensation for aid generates a sure gain rather than merely an expectation of gain.

A policy of mutual aid meets the previous section's two conditions concerning compensation. First, compensation is possible for aid given. Second, the compensation itself maximizes collective utility. It is aid received in accordance with the policy. According to a contract, assuming one exists and is costless, the one helping second compensates the one helping first. The second need not be compensated after he helps because he received his compensation in advance. The contract provides the second a large benefit now and provides the first compensation in the form of a future favor that later costs the second little. This compensation maximizes collective utility by increasing utility for the first. When custom enforces the policy, the compensation does not require keeping records, at least not in the model constructed, because the agents in the model, being rational, all comply with the policy. In an enduring group, a policy of mutual aid maximizes collective utility, with each agent providing aid in turn receiving compensation later. In ideal conditions for coordination and given a policy of mutual aid as a mechanism for compensation, collective rationality requires maximizing collective utility by implementing the policy.

14.5 Generalization

The case of mutual aid shows that collective rationality may require maximization of collective utility. Other examples arise from other schemes of compensation that maximize collective utility. This section reviews methods of compensation and computation of adequate compensation.

If one collective act has higher collective utility than another collective act, then, given the possibility of redistributing collective utility, realizing the first act may make all better off than realizing the second act. The first act achieves a higher total utility, and redistribution can make each individual's utility at least as great as under the second act. However, redistributing utility generally bears a cost, and high costs may make compensation through redistribution impractical. In a policy of mutual aid, gains in future turnabout situations provide compensation without redistribution costs. This is an advantage of the policy's method of compensation.

Some forms of compensation are inefficient and reduce collective utility, and all forms of compensation in our world have costs. To control conditions that affect the collective rationality of maximizing collective utility, I assume an idealized model in which mechanisms of compensation operate without cost.

Compensation may take many forms. Two who inherit a painting may arrange for one to buy out the other, or may take turns possessing the painting. One receives the painting, and the other receives compensation in cash or future time with the painting. A traveler may cede his reserved seat on an overbooked flight to another traveler who wants it more, and receive as compensation a travel voucher from the airline, in an exchange that on balance benefits the airline, too, and so maximizes collective utility. Compensation for the traveler giving up his seat need not generate a personal utility gain after receiving the voucher equal to the utility gain of the traveler who receives his seat. Although calculating collective utility requires interpersonal comparisons of utility, calculating adequate compensation requires only personal comparisons of utility.

The government may impose a workplace regulation that increases worker safety and health benefits at a cost to an industry that the government compensates with a tax break funded by savings from health care and disability payments and by increased revenue from income taxes that workers pay because of their greater longevity in the workforce. The industry, although it loses because of the regulation, gains on balance from other legislation. The compensation for the industry's costs comes from a third party, government, which belongs to the group with respect to which the regulation maximizes collective utility.

What is adequate compensation for an individual's loss from a collective act that maximizes collective utility? I calculate adequate compensation with respect to the alternative that would have been realized if the individual losing had not participated in the collective act, and consequently the collective act had not been realized. A collective act is an alternative to many collective acts, but the difference for an individual from the collective act that would have been realized instead constitutes adequate compensation. To motivate the individual, compensation need only be a little more than the difference. Compensation makes the result of maximizing collective utility as good for any individual as she can obtain alone. With compensation, none opposes maximizing collective utility because none does better given the alternative.

The calculation of adequate compensation for a member of a group does not presume that if the member withdraws from a collective act that maximizes collective utility the other members of the group still perform their parts in the act. The members' characters and situations settle their responses and the collective act that would be realized instead. The computation of adequate compensation depends on details of the group's collective action problem.

An individual's compensation for participation in a collective act that maximizes collective utility is with respect to the individual's position if the collective act were not realized because of the individual's failure to participate. In the Prisoner's Dilemma neither agent influences the other. If one agent acts uncooperatively, and the other agent acts cooperatively, the uncooperative agent receives her maximum payoff. Given that the total from cooperation is less than the sum of the agents' maximum payoffs, cooperation brings some agent less than her maximum payoff, and so cooperation may seem to compensate inadequately some agent despite maximizing collective utility. However, in a cooperative form of the game, where cooperation is rational, if one agent did not enter a binding contract committing each agent to cooperative behavior, then each agent would act uncooperatively, with the result that each agent receives the second lowest payoff. In the cooperative game, a contract requiring both to act cooperatively may adequately compensate each agent because the alternative to the contract brings lower total utility. Adequate compensation need only give an agent at least what she would get if she did not act cooperatively. Cooperation permits compensating each agent because it produces higher total utility than does each agent's acting uncooperatively. No agent loses by participating in cooperation, given the consequences of not acting cooperatively.

Economics introduces two types of efficiency, which I just sketch and which Chipman (2008) explains thoroughly. Each type of efficiency depends on a type of superiority. One collective act is (strictly) Pareto-superior to another if and only if the first is better for all, and it is (weakly) Pareto-efficient if and only if no alternative is (strictly) Pareto-superior to it. One collective act is Kaldor-Hicks superior to another if and only if the first achieves a state in which gainers can compensate losers without becoming losers themselves, and it is Kaldor-Hicks efficient if and only if no alternative is Kaldor-Hicks superior to it. Gains, losses, and compensation are with respect to the move from the second collective act to the first. The compensation can be in goods and need not involve interpersonal utility. It need not even involve transferable utility; transferable goods suffice. Pareto-efficiency entails Kaldor-Hicks efficiency, but Kaldor-Hicks efficiency does not entail Pareto-efficiency, because Kaldor-Hicks efficiency is just potential Pareto-efficiency. A group of shepherds achieve Kaldor-Hicks efficiency if all give their ewes to the shepherd who has the unique ram even if he does not compensate them with lambs.

The compensation schemes I consider are not just possible but executed to give individuals reasons to participate in collective acts that maximize collective utility. Adequate compensation for participation in a collective act that maximizes collective utility does not require that the act be Kaldor-Hicks efficient, but just Kaldor-Hicks superior to the collective act that would be realized in its place, if this is the same collective act given any individual's withdrawal from the maximizing act, as in bargaining problems. If the collective act that would be realized in place of the maximizing act depends on who withdraws from the maximizing act, as in coalitional games, then adequate compensation yields a collective act that is Kaldor-Hicks superior to, for each individual, the collective act that would be realized if that individual were to withdraw from the maximizing act.

14.6 Extensions

A cooperative game may include opportunities for compensation (or side payments). Then a collective act that maximizes collective utility may arise from a binding contract that provides compensation for losses. The contract may distribute the gains of cooperation to enlist the cooperation of all.

In general, a collective action problem is a game; and a combination of strategies, one for each player, such that each strategy is rational given the combination, is a (subjective) solution to the game, a rational collective act given the players' awareness of the combination. In some games, a solution maximizes collective utility. Weirich (2001: chapter 6) and Gibbard (2008: chapter 2, App.) describe bargaining games in which collective rationality requires maximizing collective utility. The previous sections describe ideal cooperative games with opportunities for compensation that maximizes collective utility and in which collective rationality similarly requires maximizing collective utility.

Support for a standard of collective rationality besides maximizing collective utility may construct games in which a solution meets the standard. A successful construction shows that the rationality of each player's act given the players' collective act ensures compliance with the standard in the games.

Besides the standard of collective-utility maximization and Pareto-efficiency, other standards of collective rationality are achieving a top collective preference and realizing a Nash equilibrium. These are goals of collective rationality that become requirements in suitable conditions. Realizing a top collective preference is a requirement in collective action problems that arise in Section 14.3's model for collective-utility maximization, because restricted to such problems collective preference has a suitable definition, according to which collective preferences among collective acts follow collective utility assignments, and realizing a top collective preference is equivalent to maximizing collective utility.

In a game, a combination of strategies, with one strategy for each player, forms an equilibrium if and only if given the combination no player has an incentive to alter her strategy. An equilibrium may not seem to be a goal of collective rationality because it is not an analog of any goal of individual rationality, except perhaps an analog of an equilibrium of desires. However, it is a goal of collective rationality because each player in the game wants to be rational given knowledge of players' choices. In ideal games of strategy in which each player knows other players' responses to her strategy, a solution is a combination of strategies such that each player's strategy maximizes personal utility given the combination. The collective rationality of the combination of strategies makes it an equilibrium. Achieving an equilibrium is a requirement of collective rationality, given players' knowledge of players' choices in ideal games of strategy.

The standards of collective rationality are consistent in cases where two or more standards apply. For example, if each member of a group rationally does her part in maximizing collective utility given that the other members do their parts, then maximizing collective utility is an equilibrium.

The Prisoner's Dilemma thus yields a lesson for a theory of rationality. It shows that extending standards of rationality from individuals to groups yields principles of collective rationality that advance goals rather than requirements of collective rationality. Ideal conditions for coordination allow a group to act efficiently, and mechanisms for compensation remove barriers to maximization of collective utility.

Bibliography

Acevedo, M. and J. Krueger (2005). "Evidential Reasoning in the Prisoner's Dilemma," *The American Journal of Psychology*, 118: 431–457.

Ahn, T.K., R.M. Isaac, and T. Salmon (2008). "Endogenous Group Formation," *Journal of Public Economic Theory*, 10: 171–194.

Ahn, T.K., M. Lee, L. Ruttan, and J. Walker (2007). "Asymmetric Payoffs in Simultaneous and Sequential Prisoner's Dilemma Games," *Public Choice*, 132: 353–366.

Alchourrón, C., P. Gärdenfors, and D. Makinson (1985). "On the Logic of Theory Change: Partial Meet Contraction and Revision Functions," *The Journal of Symbolic Logic*, 50: 510–530.

Alexander, J.M. (2014). "Evolutionary Game Theory," *Stanford Encyclopedia of Philosophy*, http://plato.stanford.edu/entries/game-evolutionary/ (accessed February 25, 2014).

Alexandrova, A. (2008). "Making Models Count," *Philosophy of Science*, 75: 383–404.

Allingham, M. (2002). *Choice Theory: A Very Short Introduction*. Oxford University Press.

Andreoni, J. (1988). "Why Free Ride? Strategies and Learning in Public Goods Experiments," *Journal of Public Economics*, 37: 291–304.

Andreoni, J. and J.H. Miller (1993). "Rational Cooperation in the Finitely Repeated Prisoner's Dilemma: Experimental Evidence," *The Economic Journal*, 103: 570–585.

Argiento, R., R. Pemantle, B. Skyrms, and S. Volkov (2009). "Learning to Signal: Analysis of a Micro-Level Reinforcement Model," *Stochastic Processes and Their Applications*, 119: 373–390.

Aristotle (1885). *Politics* (transl. Jowett). Oxford: Clarendon Press.

Arló-Costa, H. and C. Bicchieri (2007). "Knowing and Supposing in Games of Perfect Information," *Studia Logica*, 86: 353–373.

Arnold, K. and K. Zuberbühler (2006). "The Alarm-Calling System of Adult Male Putty-Nosed Monkeys, Cercopithecus Nictitans Martini," *Animal Behaviour*, 72: 643–653.

(2008). "Meaningful Call Combinations in a Non-Human Primate," *Current Biology*, 18: R202–3.

Arrow, K. (1951). *Social Choice and Individual Values*. Yale University Press.

Ashworth, J. (1980). *Trench Warfare 1914–1918*. London: MacMillan.

Aumann, R. (1987). "Correlated Equilibrium as an Expression of Bayesian Rationality," *Econometrica*, 55: 1–19.

(1995). "Backward Induction and Common Knowledge of Rationality," *Games and Economic Behavior*, 8: 6–9.

Aumann, R. and M. Maschler (1995). *Repeated Games with Incomplete Information.* MIT Press.
Axelrod, R. (1984). *The Evolution of Cooperation.* New York: Basic Books.
Axelrod, R. and W. D. Hamilton (1981). "The Evolution of Cooperation," *Science*, 211: 1390–1396.
Barclay, P. (2004). "Trustworthiness and Competitive Altruism Can Also Solve the Tragedy of the Commons," *Evolution and Human Behavior*, 25: 209–220.
Barclay, P. and R. Miller (2007). "Partner Choice Creates Competitive Altruism in Humans," *Proceedings of the Royal Society: Biological Sciences*, 274: 749–753.
Barrett, J.A. (2007). "Dynamic Partitioning and the Conventionality of Kinds," *Philosophy of Science*, 74: 527–546.
(2009). "Faithful Description and the Incommensurability of Evolved Languages," *Philosophical Studies*, 147: 123–137.
(2013a). "On the Coevolution of Basic Arithmetic Language and Knowledge," *Erkenntnis*, 78: 1025–1036.
(2013b). "On the Coevolution of Theory and Language and the Nature of Successful Inquiry," *Erkenntnis* online doi: 10.1007/s10670-013-9466-z.
Barrett, J.A. and K. Zollman (2009). "The Role of Forgetting in the Evolution and Learning of Language," *Journal of Experimental and Theoretical Artificial Intelligence*, 21: 293–309.
Battigalli, P. and G. Bonanno (1999). "Recent Results on Belief, Knowledge and the Epistemic Foundations of Game Theory," *Research in Economics*, 53: 149–225.
Battigalli, P. and M. Dufwenberg (2009). "Dynamic Psychological Games," *Journal of Economic Theory*, 144: 1–35.
Battigalli, P., A. Di Tillio and S. Dov (2013). "Strategies and Interactive Beliefs in Dynamic Games," in D. Acemoglu, M. Arellano, and E. Dekel (eds.), *Advances in Economics and Econometrics: Theory and Applications: Tenth World Congress.* Cambridge University Press.
Bayer, R. (2011). "Cooperation in Partnerships: The Role of Breakups and Reputation," *University of Adelaide School of Economics Research Paper* No. 2011-22.
Bellany, I. (2003). "Men at War: The Sources of Morale," *RUSI Journal*, 148: 58–62.
Bermúdez, J.L. (2009). *Decision Theory and Rationality.* Oxford University Press.
(2013). "Prisoner's Dilemma and Newcomb's Problem: Why Lewis's Argument Fails," *Analysis*, 73: 423–429.
Besanko, D. and R. Braeutigam (2010). *Microeconomics.* Wiley.
Bicchieri, C. (2002). "Covenants Without Swords: Group Identity, Norms, and Communication in Social Dilemmas," *Rationality and Society*, 14: 192–228.
(2006). *The Grammar of Society: The Nature and Dynamics of Social Norms.* Cambridge University Press.
Bicchieri, C. and M.S. Greene (1997). "Symmetry Arguments for Cooperation in the Prisoner's Dilemma," in G. Holström-Hintikka and R. Tuomela (eds.), *Contemporary Action Theory.* Dordrecht and Boston: Kluwer Academic Publishing.
(1999). "Symmetry Arguments for Cooperation in the Prisoner's Dilemma," in C. Bicchieri, R. Jeffrey, and B. Skyrms (eds.), *The Logic of Strategy.* Oxford University Press, pp. 175–195.

Bicchieri, C. and A. Lev-On (2007). "Computer-Mediated Communication and Cooperation in Social Dilemmas: An Experimental Analysis," *Politics Philosophy Economics*, 6: 139–168.

Bicchieri, C., A. Lev-On, and A. Chavez (2010). "The Medium or the Message? Communication Relevance and Richness in Trust Games," *Synthese*, 176: 125–147.

Binmore, K. (1991). "Bargaining and Morality," in D. Gauthier and R. Sugden (eds.), *Rationality, Justice and the Social Contract: Themes from Morals by Agreement*. New York: Harvester Wheatsheaf, pp. 131–156.

(1994). *Playing Fair: Game Theory and the Social Contract I*. Cambridge, MA: MIT Press.

(2001). "A Review of Robert Axelrod's The Complexity of Cooperation: Agent-Based Models of Competition and Collaboration," *Journal of Artificial Societies and Social Simulation* (http://jasss.soc.surrey.ac.uk/1/1/review1.html).

(2005). *Natural Justice*. Oxford University Press.

(2006). "Why Do People Cooperate?" *Politics, Philosophy & Economics*, 5(1): 81–96.

(2007a). *Playing for Real*. Oxford University Press.

(2007b). *Game Theory: A Very Short Introduction*. Oxford University Press.

(2011). *Rational Decisions*. Princeton University Press.

Binmore, K. and A. Shaked (2010). "Experimental Economics: Where Next?" *Journal of Economic Behavior and Organization*, 73: 87–100.

Board, O. (2004). "Dynamic Interactive Epistemology," *Games and Economic Behavior*, 49: 49–80.

(2006). "The Equivalence of Bayes and Causal Rationality in Games," *Theory and Decision*, 61: 1–19.

Bolton, G. and A. Ockenfels (2000). "A Theory of Equity, Reciprocity and Competition," *American Economic Review*, 90: 166–193.

Bonanno, G. (2011). "AGM Belief Revision in Dynamic Games," in R.A. Krzysztof (ed.), *Proceedings of the 13th Conference on Theoretical Aspects of Rationality and Knowledge (TARK XIII)*, ACM. New York, pp. 37–45.

(2013a). "Reasoning about Strategies and Rational Play in Dynamic Games," in J. van Benthem, S. Ghosh, and R. Verbrugge (eds.), *Modeling Strategic Reasoning, Texts in Logic and Games*. Berlin Heidelberg: Springer, forthcoming.

(2013b). "A Dynamic Epistemic Characterization of Backward Induction Without Counterfactuals," *Games and Economic Behavior*, 78: 31–43.

(2013c). "An Epistemic Characterization of Generalized Backward Induction," Working Paper No. 134, University of California Davis (http://ideas.repec.org/p/cda/wpaper/13-2.html).

Brams, S. (1975). "Newcomb's Problem and Prisoners' Dilemma," *The Journal of Conflict Resolution*, 19: 496–612.

Brandenburger, A. and A. Friedenberg (2008). "Intrinsic Correlation in Games," *Journal of Economic Theory*, 141: 28–67.

Brennan, G. and M. Brooks (2007a). "Esteem-Based Contributions and Optimality in Public Goods Supply," *Public Choice*, 130: 457–70.

(2007b). "Esteem, Norms of Participation and Public Good Supply," in P. Baake and R. Borck (eds.), *Public Economics and Public Choice: Contributions in Honor of Charles B. Blankart*. Berlin Heidelberg: Springer, pp. 63–80.

Brennan, G. and L. Lomasky (1993). *Democracy and Decision: The Pure Theory of Electoral Preference*. Cambridge University Press.

Brennan, G. and P. Pettit (2004). *The Economy of Esteem: An Essay on Civil and Political Society*. Oxford University Press.

Brennan, G. and G. Tullock (1982). "An Economic Theory of Military Tactics: Methodological Individualism at War," *Journal of Economic Behavior & Organization*, 3: 225–242.

Brennan, G., L. Eriksson, R. Goodin, and N. Southwood (2013). *Explaining Norms*. Oxford University Press.

Broome, J. (1987). "Utilitarianism and Expected Utility," *Journal of Philosophy*, 84: 405–422.

(2012). *Climate Matters: Ethics in a Warming World (Norton Global Ethics Series)*. New York: W.W. Norton & Company.

Buchanan, J.M. (1965/1999). "Ethical Rules, Expected Values, and Large Numbers," *Ethics*, 76: 1–13. Reprinted in *The Collected Works of James M. Buchanan. Vol. 1. The Logical Foundations of Constitutional Liberty*. Indianapolis: Liberty Fund Inc., pp. 311–328.

(1967/2001). "Cooperation and Conflict in Public-Goods Interaction," *Western Economic Journal*, 5: 109–121. Reprinted in *The Collected Works of James M. Buchanan. Vol. 15. Externalities and Public Expenditure Theory*. Indianapolis: Liberty Fund Inc., pp. 227–243.

(1968/1999). *The Demand and Supply of Public Goods* Chicago: Rand McNally & Co. Reprinted in *The Collected Works of James M. Buchanan. Vol. 5. The Demand and Supply of Public Goods*. Indianapolis: Liberty Fund Inc.

(1975/2000). *The Limits of Liberty: Between Anarchy and Leviathan*. University of Chicago Press. Reprinted in *The Collected Works of James M. Buchanan. Vol. 7. The Limits of Liberty: Between Anarchy and Leviathan*. Indianapolis: Liberty Fund Inc.

(1978/1999). "Markets, States, and the Extent of Moral," *American Economic Review* 68: 364–378. Reprinted in *The Collected Works of James M. Buchanan. Vol. 1. The Logical Foundations of Constitutional Liberty*. Indianapolis: Liberty Fund Inc., pp. 360–367.

(1983/2001). "Moral Community and Moral Order: The Intensive and Extensive Limits of Interaction," in H.B. Miller and W.H. Williams (eds.), *Ethics and Animals*. Clifton, NJ: Humana Press, pp. 95–102. Reprinted in *The Collected Works of James M. Buchanan. Vol. 17. Moral Science and Moral Order*. Indianapolis: Liberty Fund Inc., pp. 202–10.

Bull, M. (2012). "What is the Rational Response?" *London Review of Books*, 34, May 24, 2012.

Burke, E. (1790). *Reflections on the Revolution in France*. London: Dodsley.

Cable, D. and S. Shane (1997). "A Prisoner's Dilemma Approach to Entrepreneur–Venture Capitalist Relationships," *The Academy of Management Review*, 22: 142–176.

Camerer, C. (2003). *Behavioral Game Theory: Experiments in Strategic Interaction.* Princeton University Press.

Charness, G. and C.L. Yang (2014). "Starting Small Toward Voluntary Formation of Efficient Large Groups in Public Goods Provision," *Journal of Economic Behavior & Organization*, 102: 119–132.

Chipman, J. (2008). "Compensation Principle," in S. Durlauf and L. Blume (eds.), *The New Palgrave Dictionary of Economics.* Second Edition. London: Palgrave MacMillan.

Cialdini, R.B., R.R. Reno, and C.A. Kallgren (1991). "A Focus Theory of Normative Conduct: A Theoretical Refinement and Reevaluation of the Role of Norms in Human Behavior," in M.P. Zanna (ed.), *Advances in Experimental Social Psychology*, vol. XXIV. San Diego: Academic Press.

Cinyabuguma, M., T. Page, and L. Putterman (2005). "Cooperation Under the Threat of Expulsion in a Public Goods Experiment," *Journal of Public Economics*, 89: 1421–1435.

Clausing, T. (2004). "Belief Revision in Games of Perfect Information," *Economics and Philosophy*, 20: 89–115.

Clements, K. and D. Stephens (1995). "Testing Models of Non-kin Cooperation: Mutualism and the Prisoner's Dilemma," *Animal Behaviour*, 50: 527–535.

Climate Change: Evidence & Causes: An Overview from the Royal Society and the U.S. National Academy of Sciences, April 2014. Available at http://dels.nas.edu/resources/static-assets/exec-office-other/climate-change-full.pdf.

Corbae, D. and J. Duffy (2008). "Experiments with Network Formation," *Games and Economic Behavior*, 64: 81–120.

Croson, R., E. Fatás, and T. Neugebauer (2006). *Excludability and Contribution: A Laboratory Study in Team Production.* Mimeo, pp. 411–427.

Davenport, C. (2014). "Political Rifts Slow U.S. Effort on Climate Laws," *The New York Times*, April 15, 2014.

Davis, D.D. and C.A. Holt (1994). "Equilibrium Cooperation in Three-Person, Choice-of-Partner Games," *Games and Economic Behavior*, 7: 39–53.

Davis, D. and W.T. Perkowitz (1979). "Consequences of Responsiveness in Dyadic Interaction: Effects of Probability of Response and Proportion of Content–Related Responses on Interpersonal Attraction," *Journal of Personality and Social Psychology*, 37: 534–550.

Davis, L. H. (1977). "Prisoners, Paradox, and Rationality," *American Philosophical Quarterly*, 14: 319–327.

(1985). "Is the Symmetry Argument Valid?" in R. Campbell and L. Snowden (eds.), *Paradoxes of Rationality and Cooperation.* University of British Columbia Press, pp. 255–262.

Dawes, R.M. (1975). "Formal Models of Dilemmas in Social Decision Making," in M.F. Kaplan and S. Schwartz (eds.), *Human Judgment and Decision Processes: Formal and Mathematical Approaches.* New York: Academic Press, pp. 87–108.

DeRose, K. (2010). "The Conditionals of Deliberation," *Mind*, 119: 1–42.

Desforges, D.M., C.G. Lord, S.L. Ramsey, J.A. Mason, M.D. Van Leeuwen, S.C. West, and M.R. Lepper (1991). "Effects of Structured Cooperative Contact on

Changing Negative Attitudes toward Stigmatized Social Groups," *Journal of Personality and Social Psychology*, 60: 531–544.
Dixit, A. and S. Skeath (1999). *Games of Strategy*. London: Norton.
Dupré, J. (2001). *Human Nature and the Limits of Science*. Oxford University Press.
Eells, E. (1981). "Causality, Utility, and Decision," *Synthese*, 48: 295–329.
Ehrhart, K.M. and C. Keser (1999). *Mobility and Cooperation: On the Run (No. 99-69). Sonderforschungsbereich 504.* Universität Mannheim.
Ellickson, R.C. (1993). "Property in Land," *Yale Law Journal*, 102: 1315.
Elster, J. (2007). *Explaining Social Behavior: More Nuts and Bolts for the Social Sciences*. Cambridge University Press.
Erev, I. and A.E. Roth (1998). "Predicting How People Play Games: Reinforcement Learning in Experimental Games with Unique, Mixed Strategy Equilibria," *American Economic Review*, 88: 848–881.
Erickson, P., J. Klein, L. Daston, R. Lemov, T. Sturm, and M. D. Gordin (2013). *How Reason Almost Lost Its Mind: The Strange Career of Cold War Rationality*. Chicago University Press.
Fehl K, D.J. van der Post, and D. Semmann (2011). "Co-evolution of Behaviour and Social Network Structure Promotes Human Cooperation," *Ecology Letters*, 14: 546–551.
Fehr, E. and S. Gächter (2000). "Cooperation and Punishment in Public Goods Experiments," *American Economic Review*, 90: 980–994.
Fehr, E. and K. Schmidt (1999). "A Theory of Fairness, Competition, and Cooperation," *The Quarterly Journal of Economics*, 114: 817–868.
Felsenthal, D.S. and M. Machover (1998). *The Measurement of Voting Power: Theory and Practice, Problems and Paradoxes*. Cheltenham: Edward Elgar.
Frank, B. (1996). "The Use of Internal Games: The Case of Addiction," *Journal of Economic Psychology*, 17: 651–660.
Frohlich, N. and J.A. Oppenheimer (1970). "I Get by with a Little Help from my Friends," *World Politics*, 23: 104–120.
Frohlich, N., T. Hunt, J.A. Oppenheimer, and R.H. Wagner (1975). "Individual Contributions for Collective Goods: Alternative Models," *The Journal of Conflict Resolution*, 19: 310–329.
Gardiner, S. (2011). *A Perfect Moral Storm: The Ethical Tragedy of Climate Change*. Oxford University Press.
Gauthier, D. (1986). *Morals by Agreement*. Oxford: Clarendon Press.
(1994). "Assure and Threaten," *Ethics*, 104: 690–721.
Geanakoplos, J., D. Pearce, and E. Stacchetti (1989). "Psychological Games and Sequential Rationality," *Games and Economic Behavior*, 1: 60–79.
Gelman, A. (2008). "Methodology as Ideology: Some comments on Robert Axelrod's *The Evolution of Co-operation*," *QA-Rivista dell'Associazione Rossi-Doria*, 167–176.
Gibbard, A. (2008). *Reconciling Our Aims: In Search of Bases for Ethics*. Oxford University Press.
Gibbard, A. and Harper, W. (1978). "Counterfactuals and Two Kinds of Expected Utility," in W. Harper, R. Stalnaker, and G. Pearce (eds.), *Ifs: Conditionals, Belief, Decision, Chance, and Time*. Dordrecht: D. Reidel, pp. 153–190.

Gilboa, I. (1999). "Can Free Choice be Known?" in C. Bicchieri, R. Jeffrey, and B. Skyrms (eds.), *The Logic of Strategy*. Oxford University Press, pp. 163–174.

Gillis, J. (2013). "Climate Maverick to Retire from NASA," *The New York Times*, April 1, 2013.

(2014). "Climate Efforts Falling Short, U.N. Panel Says," *The New York Times*, April 14, 2014: 2–14.

Ginet, C. (1962). "Can the Will be Caused?" *The Philosophical Review*, 71: 49–55.

Goldman, A. 1970. *A Theory of Human Action*. Princeton University Press.

Govindan, S. and R. Wilson (2008). "Nash Equilibrium, Refinements of," in S. Durlauf and L. Blume (eds.), *The New Palgrave Dictionary of Economics*. New York: Palgrave Macmillan.

Gowa, J. (1986). "Anarchy, Egoism, and Third Images: The Evolution of Cooperation and International Relations," *International Organization*, 40: 167–186.

Gunnthorsdottir, A., D. Houser, and K. McCabe (2007). "Disposition, History and Contributions in Public Goods Experiments," *Journal of Economic Behavior and Organization*, 62: 304–315.

Güth, W., M.V. Levati, M. Sutter, and E. van der Heijden (2007). "Leading by Example with and Without Exclusion Power in Voluntary Contribution Experiments," *Journal of Public Economics*, 91: 1023–1042.

Hall L., P. Johansson, and T. Strandberg (2012). "Lifting the Veil of Morality: Choice Blindness and Attitude Reversals on a Self-Transforming Survey," *PLoS ONE* 7(9): e45457. doi: 10.1371/journal.pone.0045457.

Hall, L., P. Johansson, B. Tärning, S. Sikström, and T. Deutgen (2010). "Magic at the Marketplace: Choice Blindness for the Taste of Jam and the Smell of Tea," *Cognition*, 117: 54–61. doi: 10.1016/j.cognition.2010.06.010.

Halpern, J. (1999). "Hypothetical Knowledge and Counterfactual Reasoning," *International Journal of Game Theory*, 28: 315–330.

(2001). "Substantive Rationality and Backward Induction," *Games and Economic Behavior*, 37: 425–435.

Hamilton, C. (2013). *Earthmasters: The Dawn of the Age of Climate Engineering*. Yale University Press.

Hamilton, W.D. (1964a). "The Genetical Evolution of Social Behaviour. I," *Journal of Theoretical Biology*, 7: 1–16.

(1964b). "The Genetical Evolution of Social Behaviour. II," *Journal of Theoretical Biology*, 7: 17–52.

Hampton, J. (1986). *Hobbes and the Social Contract Tradition*. Cambridge University Press.

(1987). "Free-Rider Problems in the Production of Collective Goods," *Economics and Philosophy*, 3: 245–273.

Hardin, G. (1968). "The Tragedy of the Commons," *Science*, 162: 1243–1248.

Hargreaves-Heap, S. and Y. Varoufakis (2004). *Game Theory: A Critical Introduction*. London: Routledge.

Harper, W.L. (1988). "Causal Decision Theory and Game Theory: A Classic Argument for Equilibrium Solutions, a Defense of Weak Equilibria, and a Limitation for the Normal Form Representation," in W.L. Harper and B. Skyrms (eds.),

Causation in Decision, Belief Change, and Statistics, II. Kluwer Academic Publishers, pp. 246–266.
Hauk, E. and R. Nagel (2001). "Choice of Partners in Multiple Two-Person Prisoner's Dilemma Games: An Experimental Study," *Journal of Conflict Resolution*, 45: 770–793.
Hausman, D. (2008). "Fairness and Social Norms," *Philosophy of Science*, 75: 850–860.
(2012). *Preference, Value, Choice and Welfare*. Cambridge University Press.
Herrnstein, R.J. (1970). "On the Law of Effect," *Journal of the Experimental Analysis of Behavior*, 13: 243–266.
Hobbes, T. (1651/1928). *Leviathan, or the Matter, Forme and Power of a Commonwealth Ecclesiasticall and Civil*. Yale University Press.
Hofstadter, D. (1983). "Metamagical Themes," *Scientific American*, 248: 14–20.
Horgan, T. (1981). "Counterfactuals and Newcomb's Problem," *Journal of Philosophy*, 78: 331–356.
Houser, N. and C. Kloesel (eds.) (1992). *The Essential Peirce: Selected Philosophical Writings Volume I (1867–1893)*. Bloomington and Indianapolis: Indiana University Press.
Hume, D. (1739/1978). *A Treatise of Human Nature*. Oxford: Clarendon Press. (Edited by L.A. Selby-Bigge. Revised by P. Nidditch. First published 1739.)
Hurley, S. L. (1991). "Newcomb's Problem, Prisoners' Dilemma, and Collective Action," *Synthese*, 86: 173–196.
(1994). "A New Take from Nozick on Newcomb's Problem and Prisoners' Dilemma," *Analysis*, 54: 65–72.
Huttegger, S.M., B. Skyrms, P. Tarrès, E.O. Wagner (2014). "Some Dynamics of Signaling Games," *Proceedings of the National Academy of Science*. www.pnas.org/cgi/doi/10.1073/pnas.0709640104. Accessed March 21, 2014.
International Monetary Fund (2013). "Energy Subsidy Reform: Lessons and Implications," January. 28, 2013. Available at www.imf.org/external/np/pp/eng/2013/012813.pdf.
Isaac, R.M., D. Schmidtz, and J. Walker (1989). "The Assurance Problem in a Laboratory Market," *Public Choice*, 62: 217–236.
James, W. (1896). "A Will to Believe," *The New World, 5*. Available at www.philosophyonline.co.uk/oldsite/pages/will.ht
Jeffrey, R.C. (1983). *The Logic of Decision*. University of Chicago Press.
Jerome, K.J. (1889). *Three Men in a Boat (To Say Nothing of the Dog)*. Bristol: Arrowsmith.
Jervis, R. (1978). "Cooperation Under the Security Dilemma," *World Politics*, 30(2): 167–214.
Johansson, P., L. Hall, B. Tärning, S. Sikström, and N. Chater (2013). "Choice Blindness and Preference Change: You Will Like This Paper Better If You (Believe You) Chose to Read It!" *Journal of Behavioral Decision Making*, DOI: 10.1002/bdm.1807.
Johnson, D., P. Stopka, and J. Bell (2002). "Individual Variation Evades the Prisoner's Dilemma," *BMC Evolutionary Biology*, 2: 15.
Joyce, J.M. (2002). "Levi on Causal Decision Theory and the Possibility of Predicting One's Own Actions," *Philosophical Studies*, 110: 69–102.

Kadane, J.B. and T. Seidenfeld (1999). "Equilibrium, Common Knowledge, and Optimal Sequential Decisions," in J.B. Kadane, M.J. Schervish, and T. Seidenfeld (eds.), *Rethinking the Foundations of Statistics*. Cambridge University Press, pp. 27–46.

Kant, I. (1993). *Grounding for the Metaphysics of Morals*. Cambridge, MA: Hackett. (Ed. J.W. Ellington. First published 1785.)

Karmei, K. and L. Putterman (2013). "Play It Again: Partner Choice, Reputation Building and Learning in Restarting, Finitely-Repeated Dilemma Games," draft, Brown University.

Kavka, G.S. (1991). "Is individual Choice Less Problematic Than Collective Choice?" *Economics and Philosophy*, 7:143–165.

(1993). "Internal Prisoner's Dilemma Vindicated," *Economics and Philosophy*, 9: 171–174.

Kitcher, P. (1981). "Explanatory Unification," *Philosophy of Science*, 48: 507–531.

(1993). "The Evolution of Human Altruism," *Journal of Philosophy*, 10: 497–516.

Koertge, N. (1975). "Popper's Metaphysical Research Program for the Human Sciences," *Inquiry*, 18: 437–462.

Kolbert, E. (2014). "Rough Forecasts," *The New Yorker*, April 14, 2014.

Kreps, D.M. and R. Wilson. (1982). "Sequential Equilibria," *Econometrica*, 50: 863–894.

Krugman, P. (2014). "The Big Green Test," *The New York Times*, June 22, 2014.

Kuhn, S. (2009). "Prisoner's Dilemma," in E. Zalta (ed.), *The Stanford Encyclopedia of Philosophy* (spring 2009 edition), http://plato.stanford.edu/archives/spr2009/entries/prisoner-dilemma/.

Kuhn, T. S. (1996). *The Structure of Scientific Revolutions*. University of Chicago Press.

Ledwig, M. (2005). "The No Probabilities for Acts Principle," *Synthese*, 144: 171–180.

Ledyard, J. (1995). "Public Goods: A Survey of Experimental Research," in J. Kagel and A. Roth (eds.), *Handbook of Experimental Economics*. Princeton University Press.

Levi, I. (1986). *Hard Choices*. Cambridge University Press.

(1997). *The Covenant of Reason: Rationality and the Commitments of Thought*. Cambridge University Press.

Levitt, S. and J. List (2007). "What Do Laboratory Experiments Measuring Social Preferences Reveal about the Real World?" *Journal of Economic Perspectives*, 21: 153–174.

Lev-On, A., A. Chavez, and C. Bicchieri (2010). "Group and Dyadic Communication in Trust Games," *Rationality and Society*, 22: 37–54.

Lewis, D. (1969). *Convention*. Cambridge, MA: Harvard University Press.

(1973). *Counterfactuals*. Oxford: Basil Blackwell.

(1979). "Prisoners' Dilemma is a Newcomb Problem," *Philosophy and Public Affairs*, 8: 235–240.

(1981). "Causal Decision Theory," *Australasian Journal of Philosophy*, 59: 5–30.

Linster, B. (1992). "Evolutionary Stability in the Infinitely Repeated Prisoner's Dilemma Played by Two-state Moore Machines," *Southern Economic Journal*, 58: 880–903.

Loomis, J.L. (1959). "Communication, the Development of Trust, and Cooperative Behavior," *Human Relations*, 12: 305–315.

Luce, R.D. (1959). *Individual Choice Behavior: A Theoretical Analysis*. John Wiley and Sons.

Luce, R.D. and H. Raiffa (1957). *Games and Decisions: Introduction and Critical Survey*. John Wiley and Sons.

McAdams, R. (2009). "Beyond the Prisoner's Dilemma: Co-ordination, Game Theory, and Law," *Southern California Law Review*, 82: 209.

McClennen, E.F. (2008). "Rethinking Rationality," in B. Verbeek (ed.), *Reasons and Intentions*. Leiden: Ashgate Publishing, pp. 37–65.

McKibben, B. (1989).*The End of Nature*. London: Anchor Books.

(2001). "Some Like it Hot: Bush in the Greenhouse," *The New York Review of Books*, July 5, 2001: 38.

MacLean, D. (1983). "A Moral Requirement for Energy Policies," *Energy and the Future*, 180–197.

Mahanarayan (1888). *Urdu jest book: Lata'if-e Akbar, Hissah pahli: Birbar namah* (translated from the Urdu by F.W. Pritchett). Delhi: Matba′Jauhar-e Hind (India Office Library 1144 VT).

Maier, N.R.F. (1931). "Reasoning in Humans II: The Solution of a Problem and its Appearance in Consciousness," *Journal of Comparative Psychology*, 12: 181–194.

Maier-Rigaud F., P. Martinsson, and G. Staffiero (2010). "Ostracism and the Provision of a Public Good: Experimental Evidence," *Journal of Economic Behavior and Organization*, 73: 387–395.

Majolo, B., K. Ames, R. Brumpton, R. Garratt, K. Hall, and N. Wilson (2006). "Human Friendship Favours Cooperation in the Iterated Prisoner's Dilemma," *Behaviour*, 143: 1383–1395.

Mancuso, P. (2011). "Explanation in Mathematics," in E. Zalta (ed.), *The Stanford Encyclopedia of Philosophy* (summer 2011 edition), http://plato.stanford.edu/archives/sum2011/entries/mathematics-explanation/.

Marwell, G. and R. Ames (1981). "Economists Free Ride, Does Anyone Else?" *Journal of Public Economics*, 15: 295–310.

Michihiro, K. (2008). "Repeated Games," in S. Durlauf and L. Blume (eds.), *The New Palgrave Dictionary of Economics*. New York: Palgrave Macmillan.

Morgan, M. (2012). *The World in the Model: How Economists Work and Think*. Cambridge University Press.

Moulin, H. (2004). *Fair Division and Collective Welfare*. MIT Press.

Munger, K. and S.J. Harris (1989). "Effects of an Observer on Handwashing in a Public Restroom," *Perceptual and Motor Skills*, 69: 733–4.

Musgrave, R. and A.T. Peacock (eds.) (1958). *Classics in the Theory of Public Finance*. London: Macmillan.

Nash, J. (1950). *Non-Cooperative Games*, Princeton University. (Diss.)

(1951). "Non-Cooperative Games," *Annals of Mathematics*, 54: 286–295.

Nisbett, R. and T. Wilson (1977). "Telling More Than We Can Know: Verbal Reports on Mental Processes," *Psychological Review*, 84: 231–259.

Nordhaus, W.D. (2007). "A Review of the Stern Review on the Economics of Climate," *Journal of Economic Literature*, 45 (3): 686–702.

(2008). *A Question of Balance: Weighing the Options on Global Warming Policies*. Yale University Press.

(2013). *The Climate Casino: Risk, Uncertainty, and Economics for a Warming World*. Yale University Press.

Northcott, R. and A. Alexandrova (2009). "Progress in Economics," in D. Ross and H. Kincaid (eds.), *Oxford Handbook of Philosophy of Economics*. Oxford University Press, pp. 306–337.

(2013). "It's Just a Feeling: Why Economic Models Do Not Explain," *Journal of Economic Methodology*, 20: 262–267.

(2014). "Armchair Science," manuscript.

Nowak, M.A. and R.M. May (1992). "Evolutionary Games and Spatial Chaos," *Nature*, 359 (6398): 826–829.

(1993). "The Spatial Dilemmas of Evolution," *International Journal of Bifurcation and Chaos*, 3: 35–78.

Nowak, M.A. and K. Sigmund (1999). "Phage-lift for Game Theory," *Nature*, 398: 367–368.

(2005). "Evolution of Indirect Reciprocity," *Nature*, 437: 1291–1298.

Nozick, R. (1969). "Newcomb's Problem and Two Principles of Choice," in *Essays in Honor of Carl G. Hempel*. Dordrecht: D. Reidel.

Olson, M. (1965). *The Logic of Collective Action: Public Goods and Theory of Groups*. Harvard University Press.

Oppenheimer, J.A. (2012). *Principles of Politics: A Rational Choice Theory Guide to Politics and Social Justice*. Cambridge University Press.

Orbell, J.M. and R.M. Dawes (1993). "Social Welfare, Cooperators' Advantage, and the Option of Not Playing the Game," *American Sociological Review*, 58: 787–800.

Orbell, J.M., P. Schwartz-Shea, and R.T. Simmons (1984). "Do Cooperators Exit More Readily than Defectors?" *American Political Science Review*, 78: 147–162.

Osborne, J.M. and A. Rubinstein (1994). *A Course in Game Theory*. MIT Press.

Ostrom, E. (1990). *Governing the Commons: The Evolution of Institutions for Collective Action*. Cambridge University Press.

Ouattaraa, K., A. Lemassona, and K. Zuberbühlerd (2009). "Campbell's Monkeys Concatenate Vocalizations into Context-Specific Call Sequences," *Proceedings of the National Academy of Science*, 106(51): 22026Ð22031, doi: 10.1073/pnas.0908118106.

Parfit, D. (1984). *Reasons and Persons*. Oxford University Press.

Page, T., L. Putterman, and B. Unel (2005). "Voluntary Association in Public Goods Experiments: Reciprocity, Mimicry, and Efficiency," *Economic Journal*, 115(506): 1032–1053.

Parikh, R. (2009). "Knowledge, Games and Tales from the East," in R. Ramanujam and S. Sarukkai (eds.), *Logic and Its Applications, Third Indian Conference, ICLA 2009, Chennai, India, January 7–11, 2009. Proceedings*. Springer: Lecture Notes in Computer Science, pp. 65–76.

Paulson, H., Jr. (2014). "The Coming Climate Crash: Lessons for Climate Change in the 2008 Recession," *The New York Times*, June 21, 2014.

Peterson, M. (2006). "Indeterminate Preferences," *Philosophical Studies*, 130: 297–320.

(2009). *An Introduction to Decision Theory*. Cambridge University Press.

Pettit, P. (1988). "The Prisoner's Dilemma is an Unexploitable Newcomb Problem," *Synthese*, 76: 123–134.

Plott, C.R. (1982). "Industrial Organization Theory and Experimental Economics," *Journal of Economic Literature*, 20: 1485–1527.

Rabin, M. (1993). "Incorporating Fairness into Game Theory and Economics," *American Economic Review*, 83: 1281–1302.

Rabinowicz, W. (2000). "Backward Induction in Games: On an Attempt at Logical Reconstruction," in W. Rabinowicz (ed.), *Value and Choice: Some Common Themes in Decision Theory and Moral Philosophy*. University of Lund Philosophy Reports, pp. 243–256.

(2002). "Does Practical Deliberation Crowd Out Self-prediction?" *Erkenntnis*, 57: 91–122.

Ramsey, F.P. (1926/1931). "Truth and Probability," *The Foundations of Mathematics and Other Logical Essays*. London: Kegan Paul, pp. 156–198.

(1928/1931). "Further Considerations," *The Foundations of Mathematics and Other Logical Essays*. London: Kegan Paul, pp. 199–211.

Rand, D.G., S. Arbesman, and N.A. Christakis (2011). "Dynamic Social Networks Promote Cooperation in Humans," *Proceedings of the National Academy of Sciences, USA*, 108: 19193–19198.

Rapoport, A. (1966). *Two-Person Game Theory*. University of Michigan Press.

Reiss, J. (2012). "The Explanation Paradox," *Journal of Economic Methodology*, 19: 43–62.

Riedl, A. and A. Ule (2013). "Competition for Partners Drives Cooperation among Strangers," paper presented at the Fifth Biennial Social Dilemmas Conference, CalTech.

Riedl, A., I. Rohde, and M. Strobel (2011). "Efficient Coordination in Weakest-Link Games," IZA Discussion Paper 6223.

Risjord, M. (2005). "Reasons, Causes, and Action Explanation," *Philosophy of the Social Sciences*, 35: 294–306.

Robson, A. (1990). "Efficiency in Evolutionary Games: Darwin, Nash, and the Secret Handshake," *Journal of Theoretical Biology*, 144: 379–396.

Rodgers, D. (2011). *Age of Fracture*. Harvard University Press.

Ross, D. (2009). "Integrating the Dynamics of Multi-Scale Economic Agency," in H. Kincaid and D. Ross (eds.), *Handbook of the Philosophy of Economics*. Oxford, pp. 245–279.

Roth, A.E. and I. Erev (1995). "Learning in Extensive Form Games: Experimental Data and Simple Dynamical Models in the Immediate Term," *Games and Economic Behavior*, 8: 164–212.

Rousseau, J. (1984). *A Discourse on Inequality*, trans. M. Cranston. New York: Penguin Books.

Rubinstein, A. (2012). *Economic Fables*. Cambridge: Open Book Publishers.

Rubinstein, A. and Y. Salant (2008). "Some Thoughts on the Principle of Revealed Preference," in A. Caplin and A. Schotter (eds.), *Handbook of Economic Methodology*. Oxford University Press, pp. 116–124.

Sagoff, M. (2008). *The Economy of the Earth: Philosophy, Law, and the Environment*, second edition. Cambridge University Press.

Sally, D. (1995). "Conversation and Cooperation in Social Dilemmas: A Meta-Analysis of Experiments from 1958 to 1992," *Rationality and Society*, 7: 58–92.

Samuelson, P.A. (1954). "The Pure Theory of Public Expenditure," *Review of Economics and Statistics*, 36: 387–389.

Santos, F. C., J. M. Pacheco, and B. Skyrms (2011). "Co-evolution of Pre-Play Signaling and Cooperation," *Journal of Theoretical Biology*, 274: 30–35.

Sautter, J., L. Littvay, and B. Bearnes (2007). "A Dual-Edged Sword: Empathy and Collective Action in the Prisoner's Dilemma," *Annals of the American Academy of Political and Social Science, The Biology of Political Behavior*, 614: 154–171.

Savage, L. J. (1954). *The Foundations of Statistics*. New York: Wiley.

Scheffler, S. (2013). *Death and the Afterlife*. Oxford University Press.

Schelling, T.C. (1960). *The Strategy of Conflict*. Harvard University Press.

(1973). "Hockey Helmets, Concealed Weapons, and Daylight Saving: A Study of Binary Choices with Externalities," *The Journal of Conflict Resolution*, 17: 381–428.

Schmidtz, D. (1995). "Observing the Unobservable," in D. Little (ed.), *On the Reliability of Economic Models*. Kluwer Academic Publishers, pp. 147–172.

(2014). "Adam Smith on Freedom," in R. Hanley (ed.), *Princeton Guide To Adam Smith*. Princeton University Press.

Schmidtz, D. and E. Willott (2003). "Reinventing the Commons: An African Case Study," *University of California Davis Law Review*, 37: 203–232.

Schultz, K. (2001). *Democracy and Coercive Diplomacy*. Cambridge University Press.

Sen, A. (1977). "Rational Fools: A Critique of the Behavioural Foundations of Economic Theory," *Philosophy and Public Affairs*, 6: 317–344.

(1987). *On Ethics and Economics*. Oxford: Blackwell.

(2002). *Rationality and Freedom*. Harvard University Press.

Shackle, G.L.S. (1958). *Time in Economics*. Amsterdam: North-Holland Publishing Company.

Shapley, L.S. and M. Shubik (1954). "A Method for Evaluating the Distribution of Power in a Committee System," *The American Political Science Review*, 48 (3): 787–792.

Shick, F. (1979). "Self knowledge, Uncertainty and Choice," *British Journal for the Philosophy of Science*, 30: 235–252.

Shin, H.S. (1992). "Counterfactuals and a Theory of Equilibrium in Games," in C. Bicchieri and M.L. Dalla Chiara (eds.), *Knowledge, Belief and Strategic Interaction*. Cambridge University Press, pp. 397–413.

Skyrms, B. (1980). *Causal Necessity: A Pragmatic Investigation of the Necessity of Laws*. Yale University Press.

(1982). "Causal Decision Theory," *Journal of Philosophy*, 79: 695–711.

(2001). "The Stag Hunt." Presidential Address, Pacific Division of the American Philosophical Association. www.socsci.uci.edu/~bskyrms/bio/papers/StagHunt.pdf (accessed June 26, 2014).

(2002). "Signals, Evolution, and the Explanatory Power of Transient Information," *Philosophy of Science*, 69: 407–428.
(2004). *The Stag Hunt and the Evolution of the Social Structure*. Cambridge University Press.
(2006). "Signals," *Philosophy of Science*, 75: 489–500.
(2010). *Signals Evolution, Learning, & Information*. Oxford University Press.
Smith, A. (1759). *The Theory of Moral Sentiments*, ed. A. Millar, A. Kincaid, and J. Bell. New York: A.M. Kelley.
Sobel, J.H. (1985). "Not Every Prisoner's Dilemma is a Newcomb Problem," in R. Campbell and L. Sowden (eds.), *Paradoxes of Rationality and Cooperation*. Vancouver: University of British Columbia Press.
(1986). "Notes on Decision Theory: Old Wine in New Bottles," *Australasian Journal of Philosophy*, 64: 407–437.
Sontuoso, A. (2013). "A Dynamic Model of Belief-Dependent Conformity to Social Norms," *MPRA Paper* 53234, University Library of Munich.
Spinoza, B. de (1677/1985). *Collected Works of Spinoza*. Princeton University Press.
Spohn, W. (1977). "Where Luce and Krantz Do Really Generalize Savage's Decision Model," *Erkenntnis*, 11: 113–134.
(1999). *Strategic Rationality*, vol. XXIV of *Forschungsberichte der DFG-Forschergruppe Logik in der Philosophie*. Konstanz University.
(2003). "Dependency Equilibria and the Causal Structure of Decision and Game Situations," *Homo Oeconomicus*, 20: 195–255.
(2007). "Dependency Equilibria," *Philosophy of Science*, 74: 775–789.
(2010). "From Nash to Dependency Equilibria," in G. Bonanno, B. Löwe, and W. van der Hoek (eds.), *Logic and the Foundations of Game and Decision Theory-LOFT8, Texts in Logic and Games*. Heidelberg: Springer, pp. 135–150.
(2012). "Reversing 30 Years of Discussion: Why Causal Decision Theorists Should One-box," *Synthese*, 187: 95–122.
Stalnaker, R. (1968). "A Theory of Conditionals", in N. Rescher (ed.), *Studies in Logical Theory*. Oxford: Blackwell, pp. 98–112.
(1996). "Knowledge, Belief and Counterfactual Reasoning in Games," *Economics and Philosophy*, 12: 133–163.
Stern, N. (ed.) (2007). *The Economics of Climate Change: The Stern Review*. Cambridge University Press.
Sunstein, C. (2007). "Of Montreal and Kyoto: A Tale of Two Protocols," *Harvard Environmental Law Review*, 31: 1.
Sylwester, K. and G. Roberts (2010). "Contributors Benefit Through Reputation-Based Partner Choice in Economic Games," *Biology Letters*, 6: 659–662.
Taylor, M. (1987). *The Possibility of Cooperation*. Cambridge University Press.
(1990). "Cooperation and Rationality: Notes on the Collective Action Problem and its Solution," in K.S. Cooks and M. Levi (eds.), *The Limits of Rationality*. London: University of Chicago Press, pp. 222–240.
Taylor, M. and H. Ward (1982). "Chickens, Whales and Lumpy Goods: Alternative Models of Public Goods Provision," *Political Studies*, 30: 350–370.

Taylor, P.D. and L.B. Jonker (1978). "Evolutionary Stable Strategies and Game Dynamics," *Mathematical Biosciences*, 40: 145–156.

Tuck, R. (2008). *Free Riding*. Cambridge, MA: Harvard University Press.

Turner, P. and L. Chao (1999). "'Prisoner's Dilemma' in an RNA virus," *Nature*, 398: 441–443.

US Global Change Research Program (2014). "The National Climate Assessment," http://nca2014.globalchange.gov/

Van den Assem, M.J., D. Van Dolder, and R.H. Thaler (2012). "Split or Steal? Cooperative Behavior when the Stakes are Large," *Management Science*, 58(1): 2–20.

Waert, S. (2003). *The Discovery of Global Warming*. Harvard University Press.

Ward, H. (1990). "Three Men in a Boat, Two Must Row: An Analysis of the Three Person Chicken Pregame," *The Journal of Conflict Resolution*, 34: 371–400.

Wang, J., S. Suri, and D. Watts (2012). "Cooperation and Assortativity with Dynamic Partner Updating," *Proceedings of the National Academy of Sciences, USA*, 109: 14363–68.

Weber, R. (2006). "Managing Growth to Achieve Efficient Coordination in Large Groups," *American Economic Review*, 96: 114–126.

Weirich, P. (2001). *Decision Space: Multidimensional Utility Analysis*. Cambridge University Press.

(2010). *Collective Rationality: Equilibrium in Cooperative Games*. Oxford University Press.

(2008). "Causal Decision Theory," *Stanford Encyclopedia of Philosophy*, http://plato.stanford.edu/entries/decision-causal/.

Weisberg, M. (2012). *Simulation and Similarity: Using Models to Understand the World*. Oxford University Press.

Weitzman, M.L. (2007). "A Review of the Stern Review on the Economics of Climate Change," *Journal of Economic Literature*, 45: 703–724.

(2013). "The Odds of Disaster: An Economist's Warning on Global Warming," published on Making Sense, PBS News Hour, May 23, 2013. Available at www.pbs.org/newshour/making-sense/the-odds-of-disaster-an-econom-1/.

Wicksell, K. (1896/1958). "A New Principle of Just Taxation," in R.A. Musgrave and A.T. Peacock (eds.), *Classics in the Theory of Public Finance*. London: Macmillan and Co., pp. 72–118.

Winch, D.M. (1971). *Analytical Welfare Economics*. Harmondsworth: Penguin.

Woodward, J. (2003). *Making Things Happen: A Theory of Causal Explanation*. Oxford University Press.

Zambrano, E. (2004). "Counterfactual Reasoning and Common Knowledge of Rationality in Normal Form Games," *Topics in Theoretical Economics*, 4 (1), article 8.

Zhong, C.-B., J. Loewenstein, and K. Murnighan (2007). "Speaking the Same Language: The Cooperative Effects of Labeling in the Prisoner's Dilemma," *The Journal of Conflict Resolution*, 51: 431–456.

Index

www.ingramcontent.com/pod-product-compliance
Ingram Content Group UK Ltd.
Pitfield, Milton Keynes, MK11 3LW, UK
UKHW050112180125
453697UK00008B/152

9 781107 044357